Integration of Services into Workflow Applications

CHAPMAN & HALL/CRC
COMPUTER and INFORMATION SCIENCE SERIES

Series Editor: Sartaj Sahni

PUBLISHED TITLES

ADVERSARIAL REASONING: COMPUTATIONAL APPROACHES TO READING THE OPPONENT'S MIND
Alexander Kott and William M. McEneaney

COMPUTER-AIDED GRAPHING AND SIMULATION TOOLS FOR AUTOCAD USERS
P. A. Simionescu

DELAUNAY MESH GENERATION
Siu-Wing Cheng, Tamal Krishna Dey, and Jonathan Richard Shewchuk

DISTRIBUTED SENSOR NETWORKS, SECOND EDITION
S. Sitharama Iyengar and Richard R. Brooks

DISTRIBUTED SYSTEMS: AN ALGORITHMIC APPROACH, SECOND EDITION
Sukumar Ghosh

ENERGY-AWARE MEMORY MANAGEMENT FOR EMBEDDED MULTIMEDIA SYSTEMS: A COMPUTER-AIDED DESIGN APPROACH
Florin Balasa and Dhiraj K. Pradhan

ENERGY EFFICIENT HARDWARE-SOFTWARE CO-SYNTHESIS USING RECONFIGURABLE HARDWARE
Jingzhao Ou and Viktor K. Prasanna

FUNDAMENTALS OF NATURAL COMPUTING: BASIC CONCEPTS, ALGORITHMS, AND APPLICATIONS
Leandro Nunes de Castro

HANDBOOK OF ALGORITHMS FOR WIRELESS NETWORKING AND MOBILE COMPUTING
Azzedine Boukerche

HANDBOOK OF APPROXIMATION ALGORITHMS AND METAHEURISTICS
Teofilo F. Gonzalez

HANDBOOK OF BIOINSPIRED ALGORITHMS AND APPLICATIONS
Stephan Olariu and Albert Y. Zomaya

HANDBOOK OF COMPUTATIONAL MOLECULAR BIOLOGY
Srinivas Aluru

HANDBOOK OF DATA STRUCTURES AND APPLICATIONS
Dinesh P. Mehta and Sartaj Sahni

HANDBOOK OF DYNAMIC SYSTEM MODELING
Paul A. Fishwick

HANDBOOK OF ENERGY-AWARE AND GREEN COMPUTING
Ishfaq Ahmad and Sanjay Ranka

HANDBOOK OF PARALLEL COMPUTING: MODELS, ALGORITHMS AND APPLICATIONS
Sanguthevar Rajasekaran and John Reif

PUBLISHED TITLES CONTINUED

Integration of Services into Workflow Applications

Paweł Czarnul

Gdańsk University of Technology
Poland

CRC Press
Taylor & Francis Group
Boca Raton London New York

CRC Press is an imprint of the
Taylor & Francis Group, an **informa** business

A CHAPMAN & HALL BOOK

CRC Press
Taylor & Francis Group
6000 Broken Sound Parkway NW, Suite 300
Boca Raton, FL 33487-2742

 First issued in paperback 2019

© 2015 by Taylor & Francis Group, LLC
CRC Press is an imprint of Taylor & Francis Group, an Informa business

No claim to original U.S. Government works

ISBN-13: 978-1-4987-0646-9 (hbk)
ISBN-13: 978-0-367-37759-5 (pbk)

Visit the Taylor & Francis Web site at
http://www.taylorandfrancis.com

and the CRC Press Web site at
http://www.crcpress.com

*In memory of my father
who encouraged me to explore the world*

Contents

Preface

Service integration has become a necessity in today's distributed systems, which have in turn become very diverse and hybrid both in terms of hardware as well as software tools. The goal of this book is to describe the current state-of-the-art solutions in distributed system architectures, discuss key challenges related to integration of distributed systems, and propose solutions both in terms of theoretical aspects such as models and workflow scheduling algorithms as well as technical solutions such as software tools and APIs.

One of the key issues is integration of various types of software applications, often not designed to work together. Such a need turns up in interdisciplinary projects where software components provided by various specialists, institutions, and companies need to be integrated. The book shows how these components can be expressed as services that can subsequently be integrated into workflow applications. The latter are often described as acyclic graphs with dependencies which allow us to define complex scenarios in terms of basic tasks. As there can be many services capable of executing a given task but on various terms such as costs and performance, quality of service becomes crucial. From the point of view of a client, optimization of a global criterion with possible constraints on other quality metrics are important e.g., minimization of the workflow execution time with a bound on the cost of selected services and corresponding data processing. Optimization refers to selection of a proper service for each task and scheduling service execution at a particular point in time. Since environments can be highly dynamic in terms of changes in service availability and parameters, such optimization needs to be dynamic as well, which is also addressed in this book. Several algorithms are compared and described in terms of applicability. Advantages and limitations of the algorithms are discussed.

Furthermore, since workflow applications often process big data, the book addresses several practical problems related to data handling. This includes data partitioning for parallel processing next to service selection and scheduling, processing data in batches or streams, and constraints on data sizes that can be processed at the same time by service instances. Constraints on data handled by a workflow execution engine and workflow data footprint are discussed. It is demonstrated how these constraints and modes impact performance and costs of workflow applications in a real environment.

Optimization within a workflow execution engine is also a concern. It is possible to perform it in parallel on a multiprocessor and multicore server or even in a distributed fashion using software agents. Benefits are presented based on experiments in real environments.

The target audience of this book comprises researchers, software engineers, practitioners in parallel and distributed software, system integrators as well as tutors, teachers, and students who can benefit from the book by acquiring knowledge in these areas:

1. State-of-the-art solutions in parallel and distributed system architectures including HPC systems, SOA, multitier architectures, peer-to-peer, distributed software objects and agents, grid systems, volunteer computing, cloud computing, sky computing, and mobile computing;

2. Challenges and solutions in multidomain workflow application definition, optimization, and execution including:

   - mapping legacy software and various types of services to a generalized service model;
   - models for integration of services into workflow applications;
   - dynamic workflow scheduling algorithms and assessment of performance and suitability for various workflow applications;
   - management of data in workflow execution including constraints on capacities of a service node, workflow execution system, data caching to improve performance as well as considering storage costs in workflow application scheduling;
   - centralized, parallel and distributed workflow execution;

3. Learning from several workflow applications that were proposed, implemented, and benchmarked in a real BeesyCluster environment including scientific, multimedia, or computationally intensive workflows as well as business-type workflows. Templates were also proposed for multidisciplinary workflow applications to be used in various contexts and settings such as project planning, monitoring of facilities, online translation using mobile devices, and data analysis. These can be used by readers as templates in their work on integration of multidomain services in scientific, business, as well as multidisciplinary projects involving many partners.

I would like to thank Prof. Henryk Krawczyk, the rector of Gdansk University of Technology and researchers from the Faculty of Electronics, Telecommunications and Informatics, Gdansk University of Technology for their continuous support. Finally, it was a great pleasure to work with Randi Cohen, Senior Acquisitions Editor, Computer Science and Marsha Pronin, Project Coordinator, Editorial Project Development at CRC Press, Taylor & Francis Group.

Paweł Czarnul
Gdansk, Poland

Introduction

This book presents a concise approach to integration of loosely coupled services into workflow applications. The majority of complex tasks, especially those that span several application fields, often need to be performed by various services and parties working together. This requires modeling of a complex task in terms of simple ones, finding services capable of executing the latter, subsequent optimization, and execution. Optimization includes assignment of services to the simple tasks in such a way that predefined global and local Quality of Service (QoS) goals are optimized and satisfied with ability to react to dynamic changes of QoS terms and service failures.

The state-of-the-art in modern parallel and distributed software architectures, corresponding implementations, and APIs are analyzed in Chapter 2. This leads to identification of important challenges in service integration such as:

1. the ability to integrate systems implementing various software architectures into complex scenarios;

2. dynamic QoS monitoring;

3. dynamic data management in distributed scenarios including various storage constraints;

4. dynamic QoS optimization of complex scenarios in such systems that would be suitable for scientific, business, and interdisciplinary scenarios.

This book makes several contributions to this field.

Firstly, the book proposes a uniform concept of a *service* that can represent executable components in all major distributed software architectures used today. Mapping of software into the service concept is presented in Section 3.3. Based on that, a general dynamic optimization problem of a complex workflow application is presented in Section 3.4.2.

Secondly, dynamic monitoring of services and filtering data to be used in service evaluation and subsequent scheduling of workflow applications is presented in Section 4.1 and extended from a previous application for cloud services [65]. Depending on the nature of particular quality metrics, short-lived changes are either neglected or taken into account, either for current or past values.

Thirdly, considering the model, the author adopted the integer linear programming (ILP) method not only to select services, but to schedule execution of services installed on resources such that global QoS criteria including execution times and costs are optimized. Algorithms such as ILP, genetic, divide and conquer using ILP for subproblems and GAIN are compared, extending the work in [68]. The proposed optimization model can be applied not only for static scheduling before the workflow application is started but also in dynamic scenarios in which a workflow application needs to be rescheduled at runtime. The algorithms are described in terms of performance and various features in Section 4.3 which allows us to select one over the other for particular workflows and settings.

Furthermore, the book discusses an extended model with determination of data flows among parallel paths of a workflow application, following and extending the work in [60]. This in turn allows dynamic data parallelization in addition to selection of services in a single workflow application.

Another very important contribution of the book includes extension of the scheduling model with optimization of data management considering the following features presented in Chapter 5:

1. Activation of batch or streaming data processing modes on a per-task basis; in the latter case, particular workflow tasks may accept data packets which are processed and forwarded to following workflow nodes without waiting for all input data; subsequent data packets are processed immediately unless storage limitations appear; additionally, data packets can be processed in parallel with other data packets by various service instances.

2. Constraints on data sizes:

   • maximum size of an individual message,
   • maximum size of data handled by all instances of a particular service at the moment, and
   • storage constraints of the workflow execution system itself.

3. Data caching in order to speed up workflow execution.

Section 6.1 presents BeesyCluster [58, 69], which acts as a universal middleware that allows us to integrate distributed systems easily. It offers accounts through which users can manage their system accounts, in particular regular files, directories, and applications on distributed clusters, servers, and workstations. BeesyCluster offers single sign-on, WWW, and Web Service interfaces [69], team work environment and a host of modules that can be built on top of the middleware including version management, e-commerce including service auctioning, wikis etc. Compared to other middleware such as Globus Toolkit, Unicore etc., it offers the following combination of features:

1. Attaching various systems quickly as the systems are accessible through SSH and require only registration of proper IP addresses and security keys.

2. Ease of addition of user accounts as only credentials need to be stored.

3. Publishing applications that act as services instantly and granting rights to other users including making use of the embedded virtual payment system. This applies to services published from various systems including commodity servers and workstations as well as even mobile devices [54].

The same chapter discusses a workflow management system built on top of the BeesyCluster middleware. Compared to other workflow management systems dedicated to certain types of applications such as grids [222] or clouds [123, 34, 116], the proposed solution makes it possible to integrate services, either your own or made available by others, into workflow applications according to the previously proposed models and subsequent optimization including workflow scheduling, execution, and monitoring. The workflow management system includes the previously proposed or extended algorithms. These can work statically before the workflow application starts or dynamically, which activates rescheduling at runtime. The latter is performed when:

1. new services appear,

2. one of the previously chosen services has become unavailable, or

3. QoS parameters of the services have changed.

In terms of workflow execution engines, the following solutions are described:

1. a centralized engine implemented by the author within a Java EE server,

2. a distributed execution engine outside of BeesyCluster using JADE [42] software agents for management of workflow execution, and

3. a concept of distributed execution using multiple application servers.

For the first two solutions, performance results were provided showing how the solutions scale when increasing sizes of resources such as the number of CPU cores, CPUs, or the number of managing nodes.

Furthermore, Chapter 7 presents several implemented real workflow applications, corresponding results, and typical performance dependencies. Examples include:

1. workflow applications for multimedia for processing of RAW images, numerical computations, and for data analysis, and

2. a busines- type workflow for purchasing of components, integration, and subsequent distribution.

The chapter also presents several concepts. The first one is a workflow application for planning interdisciplinary projects that may involve both scientific and business-type services. Furthermore, modeling hybrid scientific and business workflow applications is presented including monitoring facilities, traffic optimization with services connected to mobile devices and HPC services, online translation, and monitoring of objects.

These workflow applications can be used as templates or recommendations for readers interested in scientific, business, or workflows integrating services from various domains. Patterns that are used in these workflow applications were outlined. Presented results and conclusions allow us to assess expected performance or issues. The presented contributions address many important aspects in the cycle of development, optimization, and execution of workflow applications applicable to multiple applications, especially spanning several fields of study.

About the Author

Paweł Czarnul is an Assistant Professor at the Department of Computer Architecture, Faculty of Electronics, Telecommunications and Informatics, Gdansk University of Technology, Poland. He obtained his MSc and PhD degrees from Gdansk University of Technology in 1999 and 2003 respectively. He spent 2001-2002 at the Electrical Engineering and Computer Science Department of University of Michigan, U.S.A. working on code parallelization using dynamic repartitioning techniques. He is the author of over 65 publications in the area of parallel and distributed computing, including journals such as International Journal of High Performance Computing Applications, Journal of Supercomputing, Metrology and Measurement Systems, Multiagent and Grid Systems, Scalable Computing: Practice and Experience etc. and conferences such as Parallel Processing and Applied Mathematics (PPAM), International Conference on Computational Science (ICCS), International Conference on Distributed Computing and Networking (ICDCN), International Multiconference on Computer Science and Information Technology (IMCSIT) and others. Dr. Czarnul actively participated in 13 international and national projects including those for industry, being the principal investigator of 3 projects. He is a reviewer for international journals such as Journal of Parallel and Distributed Systems, Journal of Supercomputing, Scientific Programming Journal, Computing and Informatics, Journal of Communication Networks and Distributed Systems etc., and conferences including PPAM, ICCS, Workshops on Software Services, Workshop on Cloud-enabled Business Process Management, International Baltic Conference on Databases and Information Systems (BalticDB&IS), Intelligent Data Acquisition and Advanced Computing Systems (IDAACS), International Conference on Information Technology (ICIT) and others. He is an expert in an international board on Software Services within the SPRERS project. He teaches the architectures of Internet systems, high performance computing systems, parallel programming, and parallel and distributed algorithms.

His research interests include parallel and distributed computing including service oriented approaches, high performance computing including accelerators and coprocessors, Internet and mobile technologies. He is leader of the BeesyCluster project that provides a middleware for a service oriented integration of computational and general purpose hardware resources with an advanced workflow management system; a co-designer of the Comcute project for volunteer computing within user browsers with control of redundancy and reliability; and a co-designer of KernelHive for automatic parallelization of computations among clusters with GPUs and multicore CPUs. Dr. Czarnul is author of DAMPVM/DAC for automatic parallelization of divide-and-conquer applications.

List of Figures

List of Tables

List of Listings

Symbol Description

T	a set of workflow tasks
E	a set of edges that determine dependencies between tasks
t_i	task i of the workflow application
S_i	a set of alternative services, each of which can perform functions required by task t_i
s_{ij}	the j-th service that is capable of executing task t_i
c_{ij}	the cost of processing a unit of data by service s_{ij}
P_{ij}	provider who has published service s_{ij}
N_{ij}	node (cluster/server) on which service s_{ij} was installed
sp_N	speed/performance of node N
d_i	size of data processed by task t_i
d^{out}	size of output data from task t_i
$f_{t_i}()$	a function that determines the size of output data based on the size of input data: $d_i^{\text{out}} = f_{t_i}(d_i^{\text{in}})$
d_{ij}^{in}	size of input data processed by service s_{ij}
d_{ij}^{out}	size of output data produced by service s_{ij}
d_{ijkl}	size of data passed from service s_{ij} to service s_{kl}
$t_{ij}^{\text{exec}}(d_{ij}^{\text{in}})$	execution time of service s_{ij}
t_{ijkl}^{comm}	communication time corresponding to passing of data of size d_{ijkl} from service s_{ij} to service s_{kl}
$t_{N_{ij}N_{kl}}^{\text{startup}}$	startup time corresponding to a communication link between nodes N_{ij} and N_{kl}
$bandwidth_{N_{ij}N_{kl}}$	bandwidth corresponding to a communication link between nodes N_{ij} and N_{kl}
t_{ijkl}^{tr}	additional translation time required if output data from service s_{ij} is sent as input data to service s_{kl} if providers of the services are different $P_{kl} \neq P_{ij}$
t_i^{st}	starting time for service $s_{i\ sel(i)}$ selected to execute t_i
t_{workflow}	wall time for the workflow i.e., the time when the last service finishes processing the last data
B	budget available for execution of a workflow
$Succ(t_i)$	a set of tasks that follow task t_i in the workflow graph
ds_{ij}^{max}	the maximum storage capacity of the machine on which service s_{ij} is installed and where it runs
d_{input}	size of the input data that is passed to the workflow
$WEC(t)$	the maximum size of data that can be held at time t in the storage spaced used by

the workflow execution engine

$SDP_{i;k}$ the size of a data packet that is sent between tasks t_i and t_k

$PARDS_{dl;i}$ the number of parallel streams used for passing data from initial data location dl to the service assigned to task t_i of the workflow application

$PARDS_{i;k}$ the number of parallel streams used for passing data between services selected for execution of tasks t_i and t_k such that $(t_i, t_k) \in E$

B_{WEC} budget allocated for the cost of storage used by the workflow execution system

AF task a task for which all output data is passed to each successor: t_i: $\forall_j \forall_{k:t_k \in Succ(t_i)} d_{ij}^{out} = \sum_l d_{ijkl}$,

DF task a task for which output data is partitioned among successors: t_i: $\forall_j d_{ij}^{out} = \sum_{l,k:(v_i,v_k)\in E} d_{ijkl}$

$d_i^p(t)$ data packets processed for task t_i at particular moment t (might be in various stages of processing)

DF(A,t) the total workflow footprint of workflow application A at time t

Chapter 1

Understanding the Need for Service Integration

1.1   Introduction

In today's information systems, integration and efficient cooperation among various types of services is a crucial challenge. Apart from clusters, servers, desktop computers, there is a growing number of mobile devices including laptops, netbooks, tablets, smartphones, as well as devices embedded in our houses and vehicles.

Mobile and embedded devices feature a variety of sensors and input devices such as ambient light sensors (ALS), accelerometers, Global Positioning System (GPS) sensors, compasses, gyroscopes. Some modern devices also feature [85] heart monitors, altimeters, temperature, humidity, perspiration sensors, cameras, microphones, etc. All of these provide streams of data that need to be processed efficiently. In certain applications such as runtime language translation combined with image recognition or identification of faces, buildings [228], etc., the latter require powerful HPC systems for processing. Fortunately, there are modern powerful parallel systems [19] that allow processing of such large data sets. Compute devices include multicore processors, accelerators with several tens of cores such as Intel Xeon Phi coprocessors, GPU cards, clusters of nodes equipped with accelerators and interconnected with fast networks such as InfiniBand, etc. It is now more than ever that cooperation of distributed input devices such as cameras and sensors as well as HPC systems is so important. There are several applications for those such as:

- Recognition of buildings or landmarks [228]

1

- Security—monitoring public facilities and places such as factories, stadiums, streets etc. [74]

- Safety—monitoring roads, traffic [154, 158], detection of collision between vehicles etc.

- Entertainment—recognition of music, systems such as AudioTag [3]

- Science—processing data from radiotelescopes [160], [181]

- Business—providing location-based services to users such as local radio and video streaming, nearby restaurants, shopping services such as informing users about current offers near the point of purchase [25]

1.2   Software Systems and Applications

However, such integration is far from easy. One of the reasons for this fact is that software architectures and implementations were developed with particular types of applications in mind. For example:

- Shared memory and distributed memory models for HPC systems; applications include demanding processing in fields such as electromagnetics [80, 190, 191], medicine [73], weather forecasting, image processing [70]. Implementations include OpenMP [49], Pthreads, Java threads [47] for shared memory machines, NVIDIA CUDA [188], OpenCL [129, 188] and OpenACC [217, 178] for parallel programming on GPU cards, OpenCL for hybrid programming on multicore CPUs and GPUs, MPI [218] for both distributed and shared memory machines (model with processes and support for threads), grid versions of MPI such as MPICH-G2 [126], PACX-MPI [128], and BC-MPI [59] for parallel programming on several clusters at the same time.

- Client-server [166] commonly used in communication on the Internet, e.g., using HTTP(S) protocols, suitable for serving general requests from clients including Web browsers which can also be extended for higher-level protocols.

- Distributed object models such as Common Object Request Broker Architecture (CORBA) [115, 164, 194] that allows cooperation among objects in a distributed system [55, 56]. Assessment of CORBA and Web Services for distributed applications is presented in [31].

- Service Oriented Architecture (SOA) [185] in which loosely coupled services can be invoked separately or integrated into more complex tasks

[144]; Web Services [185] as implementation of SOA have gained popularity as a way to automate cooperation between businesses. The standards and protocols such as WSDL, SOAP and Universal Description, Discovery and Integration (UDDI) form the basis of Web Services. Recently, REST style Web Services [180] have gained attention as a light weight solution for invoking services. While the SOAP style Web Service might use (although not necessarily) HTTP(S), REST services rely on mapping onto HTTP methods.

- Agent systems—in distributed agent systems [42], agents are autonomous entities that can cooperate either to perform a common goal collectively, e.g., in a distant part of the network, or that can be launched for negotiation, trade or monitoring. The key element is intelligence and autonomy of agents. Examples of popular environments include JADE [42], which follows FIPA standards.

- Volunteer computing in which Internet clients donate parts of their computing resources such as processing power of CPUs and GPUs to process data packets and produce partial results of large-scale computations. This allows either data parallelization for problems that rely on independent computations of various data sets or processing multiple cases or scenarios by individual Internet clients. The server side of the system must ensure that results returned are correct by verification against results of other clients. Examples of such systems include BOINC [162, 30], Comcute [6, 75, 62] and more recently a new WeevilScout prototype framework [52] for engaging volunteers using Internet browsers with JavaScript for computations.

- Grid computing integrates distributed Virtual Organizations (VOs) and can be thought of as controlled resource sharing. Independent institutions (VOs) governed by various policies and rules related to access to resources, security measures, resource management tools, queuing systems on clusters, etc., are integrated into a single high-level system. This can be done with the help of so-called *grid middleware* such as Globus Toolkit, GridBus or UNICORE. On top of the latter, high level services, APIs, and GUIs can be built to manage a distributed environment, often spanning institutions in several countries. The goal is to give the user an impression of working in a single uniform environment by hiding all differences of the VOs.

- Cloud [29] and sky computing [127] outsourcing one of the following: infrastructure (Infrastructure as a Service—IaaS), software (Software as a Service—SaaS) or whole hardware-software platforms (Platform as a Service—PaaS). The user relies on a provider to support the given service including necessary updates, security of data, and maintaining constant and satisfactory agreement. Data security and the so-called vendor lock-in issue are the key concerns of the client in this processing model. Cloud computing is marketed by providers as more cost effective than maintaining owned software and hardware. Sky computing can be thought of as parallel utilization of many independent clouds as an attempt to address those concerns.

1.3   Rationale behind Service Integration

By definition, each of the aforementioned parallel or distributed software architectures was designed to integrate components with a particular application type in mind. Invocation of a remote software component always involves at least the following steps:

1. discovery of a peer (requires an identifier of another party),

2. invocation and communication (requires data representation format and a protocol for data delivery).

Depending on the main focus and importance such as high performance (HPC systems but also volunteer-based and grid systems), interoperability among components within a system (distributed object systems), remote access to third-party services (client-server, SOA, grid systems, clouds), distributed monitoring, data acquisition, cooperation (agent systems), various means of discovery, invocation and communication with a remote side are used. Table 1.1 lists details for exemplary implementations of software architectures.

TABLE 1.1: Software architecture exemplary implementations: discovery and communication.

Software architecture/system type	Exemplary implementation	Discovery/ identifier	Communication/ protocols
HPC system (distributed memory)	MPI [107, 108]	Accessible through an id (process rank)	Message sent through MPI runtime, can use shared memory, Infiniband, etc.

Distributed object system	CORBA [164] implementation	Interoperable Object Reference (IOR)	GIOP implementations, e.g., Internet InterORB Protocol (IIOP) that specifies that Object Request Brokers (ORBs) send messages over TCP/IP [199] for communication on the Internet
Client-server	Browser/Java servlet [117], browser/PHP [159] etc.	URL	HTTP(S)
Client-server	BSD stream sockets [199]	<host name >:<port number>	TCP/IP [199]
Client-server	Enterprise Java Beans [183]	Through JNDI [168]	RMI, e.g., RMI over IIOP
SOA [88]	SOAP Web Services [88]	UDDI/service registry for finding a WSDL of a service, service available at an URL	SOAP + e.g., HTTP(S)
Agent system	JADE [42]	Can use a Directory Facilitator (DF) that allows agents to publish services which enable others to discover agents offering specific functions, communication among agents uses an AgentID that identifies an agent	Message passing using ACLMessage (Agent Communication Language), uses FIPA message types such as QUERY, PROPOSE, INFORM, specifies an ontology and a language for message content
Grid system/middleware	Globus Toolkit's GRAM [206] which is a uniform interface for submission of jobs in grid systems	Service's contact address uses form grid.domain:<port>/jobmanager-sge:/C=US/ O=Example/OU=Grid/ CN=host/grid.domain	GRAM service listening on a specified port, GRAM uses Grid Security Infrastructure for security, a proxy certificate needs to be obtained before job submission

Cloud system	Amazon S3 [156] for cloud storage	REST API/URL [156]	HTTP, a valid AWS Access Key ID in order to authenticate requests for storing buckets

In today's information systems, which are more and more heterogeneous and diverse, the ways and possibilities designed within the aforementioned approaches are not always sufficient. This is because there is a growing need for cooperation of various systems in various fields of study, for example:

1. Gathering various types of data from many sensors located in both fixed locations, mobile stations, and even mobile devices—this may refer to monitoring pollution, noise, dangerous events in a city like traffic rule violations, fights, monitoring sports events, acceleration and location of devices in vehicles like trains, planes, buses, cars, people falling down, and many others. These generate many streams of data that need to be processed for detection of various situations. This requires coupling the streams with either dedicated powerful HPC systems or resources installed within grid- or cloud-based systems (that may include HPC solutions as well).

2. Integration of various types of services in a single undertaking, e.g., design of an airport, coupling (at various time points) social polls, decisions taken by local authorities, analysis regarding environmental impact, design of the infrastructure, air traffic simulations, weather analysis and forecasts, etc.

3. Analysis of blogs, websites, posts on various websites can be performed both by dedicated engines on the websites but also delegated to other standalone HPC facilities.

In general, integration of software services into a workflow application requires the following steps depicted in Figure 1.1:

1. Definition of a complex task—this can usually be done as a workflow (application) represented by a directed graph in which nodes correspond to basic tasks and edges denote dependencies between connected tasks.

2. Searching for software services capable of performing given basic tasks—basic tasks need to accept formats of output data produced by preceding services and produce output understood by successors.

3. Optimization of workflow scheduling, i.e., selection of a service for each task and setting a time slot for its execution such that a certain optimization criterion is minimized, e.g., the execution time, with possibly other constraints imposed on the total cost of services, etc. Static scheduling occurs before the workflow application is to be executed.

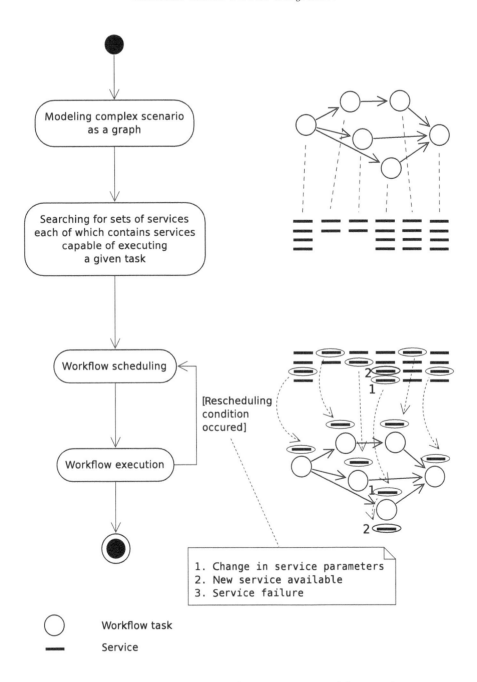

FIGURE 1.1: Steps in integration of services into workflow applications.

4. Workflow execution and monitoring involves actual invocation of services selected for the tasks, passing of data between services, and monitoring of workflow instance state. Furthermore, if there has been a change in service parameters, a previously selected service has failed or new services have appeared, rescheduling may be launched in order to improve the current schedule.

1.4   Challenges in Service Integration

Integration of various software components is becoming increasingly difficult and faces the following challenges, only partially addressed in current systems:

1. Service discovery. In the process shown in Figure 1.1 sets of services need to be assigned to particular workflow tasks (nodes). This can be done by a workflow creator who can browse service registries or known locations for services that are capable of performing the given task. It is important to note that the services must accept data in a specified format and produce output in a desired format so that services assigned to various tasks can cooperate. Alternatively, services can be found using automated queries assuming functional specification of required and available services are expressed in a common language.

2. Establishing a common middleware that would allow hiding details of various types of software irrespective of business applications realized by this software.

3. Setting up a common Quality of Service (QoS) definition for various types of services.

4. Incorporation of human activities into the process, along software services.

5. Handling data sets in various ways, even in a single workflow application:

   - batch or stream processing,
   - large data sets with constraints on maximum data sizes that can be handled by individual service instances, resources on which services are installed, and the workflow execution system.

6. Development of an integrated software platform supporting the above for various types of software and management of several instances of such integrated projects at the same time.

1.5   Solutions Proposed in the Book

The goal of this book is the proposal of a solution to the integration problem stated above with management of quality including optimization of execution in distributed environments. This is fundamentally possible by application
of the following concepts:

1. *Service*—the book presents how existing legacy software can be converted and exposed as services as well as how various types of existing
   services can be uniformly described and used; the concept of the service
   that is introduced allows generalization at two levels:

   (a) Technology—various types of services are defined within a unified
       concept of a service; these include:
       - legacy software
       - Web services
       - grid services
       - cloud services

   (b) Application—services from the following fields can now be de-
       scribed using the same ontology:
       - Business—oriented more on realizing tasks in e-commerce. In-
         tegration of such services is more control flow oriented.
       - Scientific—performing usually compute intensive tasks and can
         operate on data sets of huge sizes.
       - Ubiquitous—services that are usually executed within a spe-
         cific context, e.g., after particular conditions have occurred.
       - Human—human activities should be integrated along with the
         aforementioned types of services.

2. *A uniform QoS model for all types of services*—all the aforementioned
   types of services can be described using one consistent QoS model.

3. *Integration of various services into workflow applications*—so far workflow application management has been considered within particular
   fields of study such as business or scientific with integration of corresponding types of services running on top of particular middleware or
   systems such as grids or clouds. Interoperability was possible within
   limited types of systems such as clouds, grids, and clusters. The book
   proposes a general approach to integration of even more types of services
   running in various systems into a single workflow application, thanks to
   the uniform service model and description.

4. *Definition of streaming and batch processing modes in a single workflow.*

5. *Consideration of various data constraints*: Processing large amounts of data from potentially various geographically distributed sources by distributed services to accomplish a given goal. This process has to consider local memory constraints, buffers, and bottlenecks caused by these.

6. *Exchangeable optimization algorithms*: A design and implementation capable of plugging in various optimization algorithms with two types of optimization:

   - Static—before workflow execution.
   - Dynamic—during workflow execution when one of the following conditions have occurred:

     (a) some previously selected service or services have failed,
     (b) QoS conditions of the services have changed, or
     (c) new interesting offers have appeared.

7. *Demonstration of scalability for real applications* in terms of processor cores within a node as well as multiple nodes of a cluster or a distributed system.

The outline of the book is as follows.

Chapter 2 discusses existing parallel and distributed systems including state-of-the-art architectures and technologies. In particular, Section 2.1 presents APIs and software middleware for parallel and distributed systems including for HPC systems Service Oriented Architectures, multitier distributed systems, peer-to-peer, grid computing, volunteer-based systems, cloud and sky computing, and mobile computing. Section 2.2 describes models and algorithms used in workflow management systems. In view of the discussed state-of-the-art in parallel and distributed systems, Section 2.3 formulates challenges and subsequently solutions to the latter proposed by the author. In particular, Section 2.3.1 shows how to integrate systems using various software architectures together. Section 2.3.2 discusses integration of services for various target applications with dynamic monitoring and evaluation of services in Section 2.3.3, data management in Section 2.3.4 and dynamic optimization in Section 2.3.5.

Chapter 3 formulates the concept of dynamic quality management for service-based workflow applications in distributed environments. Section 3.1 describes a generalized model of a service. Section 3.2 follows with a generalized model of a distributed system applicable to the distinguished types of systems. Section 3.3 presents a proposed mapping of various types of modern software to a generalized service. Section 3.4 presents a solution for dynamic QoS-aware optimization of service-based workflow applications. This is described in terms of the architecture (Section 3.4.1) and the workflow application scheduling model (Section 3.4.2) followed by definition of a QoS-aware workflow scheduling problem in a static version (Section 3.4.3), dynamic (Section 3.4.4), and dynamic with data streaming (Section 3.4.5).

In Chapter 4 the author discusses scheduling algorithms for the proposed generalized mapping of workflow applications to distributed systems. Section 4.1 discusses how important quality metrics should be monitored (Section 4.1.1) and filtered (Section 4.1.2). Section 4.2.1 presents a workflow scheduling algorithm using integer linear programming (ILP), Section 4.2.2 presents application of a genetic algorithm, Section 4.2.3 a divide-and-conquer approach, and in Section 4.2.4 a hill-climbing type of algorithm based on iterative substitution of a single service. Section 4.3 presents characteristics and comparison of algorithms while Section 4.4 presents extensions to the model allowing automatic data parallelization, and Section 4.5, modeling additional constraints. Section 4.6 suggests how checkpointing can be used to optimize workflow application execution.

Chapter 5 presents solutions for dynamic data management in workflow execution. Section 5.1 extends the previously proposed workflow scheduling model with multithreading (Section 5.2), data streaming for particular workflow tasks (Section 5.3), and constraints for intermediate data storage during workflow execution (Section 5.4). The impact of the message size on the workflow execution size is investigated in Section 5.4.1, impact of the storage size in 5.4.2, and the storage constraints for particular services in Section 5.4.3. Importance of caching during workflow execution is shown in Section 5.4.4 and the total workflow data footprint in Section 5.4.5. Cost of storage in the context of workflow execution is discussed in Section 5.4.6. Section 5.5 considers the importance of coherent interfaces for data exchange between services in workflow applications.

Chapter 6 presents a middleware for both publishing and consuming services offered by various independent providers as well as a workflow management system that uses the former. Section 6.1 presents BeesyCluster, a platform proposed by the author and subsequently co-designed and co-developed. BeesyCluster was developed as a middleware for integration of distributed services compliant with the requirements presented in Section 6.1.1. The domain object model is shown in Section 6.1.2 while BeesyCluster's functions and features as middleware are presented in Sections 6.1.3 and 6.1.4. Section 6.1.5 discusses the internal architecture of BeesyCluster with the most important components. Section 6.1.6 presents adopted security solutions. Management of services on the platform is shown in Section 6.1.7. Furthermore, two interfaces for clients are presented: WWW in Section 6.1.7.1 and the Web Service interface in Section 6.1.7.2. Means to publish services and assign privileges to other users are shown in Section 6.1.8, while Section 6.1.9 presents efficiency of the interfaces for the most typical user commands. Section 6.2 presents a workflow management system designed and implemented by the author on top of the BeesyCluster middleware. Section 6.2.1.1 presents functional requirements for the system, and Section 6.2.1.2 covers analysis and design of modules such as workflow editor, execution engine, and key components. Workflow execution is described in Section 6.2.2 with a centralized version in Section 6.2.2.1, distributed using software agents in Section 6.2.2.2, and distribution using multiple application servers in Section 6.2.2.3.

Chapter 7 presents various practical workflow applications modeled, designed and executed using the proposed workflow model and management system in BeesyCluster. Section 7.1 describes selected patterns that can be expressed using the proposed workflow models. Section 7.2 presents a way to model and optimize scientific and multimedia applications, Section 7.3 a business-oriented production problem. Section 7.4 presents a template for a workflow application for planning interdisciplinary projects. Controlling distributed devices and sensors is shown in Section 7.5 with monitoring facilities in Section 7.5.1, integration of resources for optimization of traffic and emergency in Section 7.5.2, a workflow application for online translation in Section 7.5.3, and monitoring distributed objects in Section 7.5.4. Finally, Section 7.6 presents a workflow application for parallel processing of data sets. Modeling performance of applications in BeesyCluster's workflow management system is discussed in Section 7.7.

Chapter 2

Existing Parallel and Distributed Systems, Challenges, and Solutions

2.1   Modern Software Architectures, Systems, and APIs for Parallel and Distributed Systems

This chapter presents the most important of modern distributed software architectures by outlining key components and the way the latter interact and synchronize. Correspondingly, examples of representative implementations of these architectures are given with discussions on the APIs available to the client. Secondly, workflow management systems for various environments are shown as a feasible way of integrating software components within particular types of systems. This is done as a prerequisite to analysis and discussion of future directions, challenges in *dynamic* integration of *various* types of distributed software to which the author proposes solutions covered in depth further in this book.

2.1.1   HPC Systems

High-performance computing [46, 47] has come a long way from dedicated, extremely expensive supercomputers to machines equipped with multicore CPUs and powerful GPU devices widely available even in our homes today. As an example, Intel Xeon Processor E7-8890 v2 features 15 cores, Intel Xeon Phi 7120 features 61 cores. NVIDIA GeForce GTX Titan features 2688 CUDA cores. AMD Opteron 6380 Series Processors feature 16 cores. Figures 2.1 and 2.2 present the performance in Tflop/s and the number of cores of the first cluster on the TOP500 [19] list over the last years. Figure 2.3 shows the average processor frequency used in the ten most powerful clusters on the TOP500 list in respective years. It can be seen that the growth in the clock frequency has practically stopped and consequently the increase in the performance is mainly due to a growing number of processor cores in the cluster as well as adoption of new computing devices. Apparently, this will also be the tendency of cheaper machines which can accommodate GPU devices easily assuming a powerful enough power supply is installed.

Adoption of more computational devices (such as computers, cluster nodes but also cores within devices) has crucial consequences for software. In order to use the available processing capabilities, one or more of the following techniques needs to be adopted:

1. launching more applications with processes/threads to use the available processors and cores,

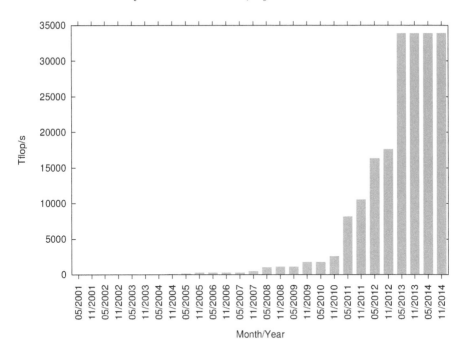

FIGURE 2.1: Performance of the first cluster on the TOP500 list, based on data from [19].

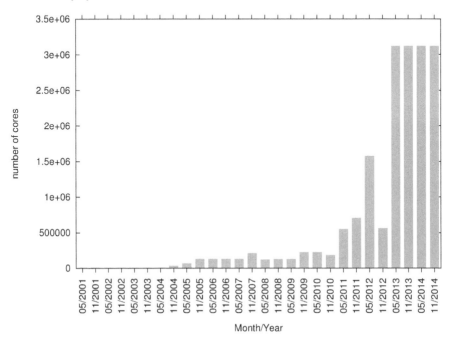

FIGURE 2.2: Number of processing cores of the first cluster on the TOP500 list, based on data from [19].

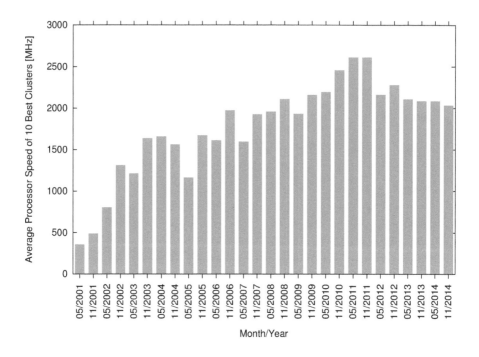

FIGURE 2.3: Average processor speed of the ten best clusters on the TOP500 list, based on data from [19].

2. parallelize existing applications so that these launch more threads to make use of the large numbers of cores.

In order to do this, a proper API and runtime systems are needed. The most popular ones are discussed next along with access and management of the parallel software.

2.1.1.1 GPGPU and Accelerators

NVIDIA CUDA [129, 188, 161] and OpenCL [38] allow General-Purpose computing on Graphics Processor Units (GPGPU). NVIDIA CUDA offers an API that allows programming GPU devices in the SIMT fashion. The following concepts are distinguished regarding the application paradigm:

1. a 3D grid that is composed of blocks aligned in three dimensions X, Y, and Z, in which

2. each block consists of threads aligned in dimensions X, Y, and Z.

The usual steps in application execution include the following:

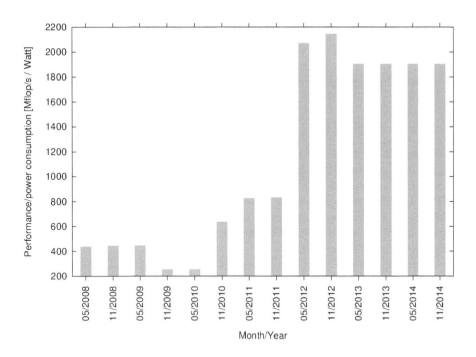

FIGURE 2.4: Performance per Watt of the first cluster on the TOP500 list, based on data from [19].

1. discovery of available GPU devices;

2. uploading input data from the RAM memory of the computer into the global memory of the device;

3. specification of grid sizes, both in terms of block counts in three dimensions as well as thread counts in three dimensions within a block;

4. launching a kernel, which is a function that is executed in parallel by various threads that process data; and

5. downloading output data produced by the kernel from the global memory of the GPU to the RAM.

Advanced code will use several optimization techniques such as:

- data prefetching along with usage of faster but smaller shared memory as well as registers;

- memory coalescing, which speeds up access to global memory;

- optimization of shared memory and register usage, which may increase the parallelization level;

- loop unrolling;

- launching several kernels and overlapping communication and computations;

- using multiple GPUs.

OpenCL offers an API similar to that of NVIDIA CUDA but generalized to run not only on GPUs but also on multicore CPUs. This allows us to use a modern workstation with several multicore CPUs and one or possibly more GPUs as a cluster of processing cores for highly parallel multithreaded codes. Because of that, OpenCL requires some more management code related to device discovery and handling but the kernel and grid concepts have remained analogous to NVIDIA CUDA.

Development and execution of both NVIDIA CUDA- and OpenCL-based applications involve the following:

1. logging into the computer equipped with CPU(s) and GPU(s), e.g., using SSH,

2. compilation of the code using a proper compiler, e.g., `nvcc` for NVIDIA CUDA and `nvcc -lOpenCL` for OpenCL programs, and

3. execution of the program.

As explained above, the aforementioned solutions require access to HPC resources using, e.g., SSH as well as programming and operating system knowledge to develop and run corresponding applications.

OpenACC [178, 217], on the other hand, allows us to extend sequential code using directives and functions that allow subsequent parallelization of the application when running on GPUs. This approach is aimed at making parallel programming on GPUs easier similarly to parallelization using OpenMP [210, 49].

Intel Xeon Phi coprocessors allow us to use one of a few programming APIs such as OpenMP [210, 49], OpenCL [38] or Message Passing Interface (MPI) [36] described below.

2.1.1.2   Clusters

Traditionally, the Message Passing Interface (MPI) [36, 218] has been used for multiprocess, multithreaded programming on clusters of machines, each of which may use multiple, multicore CPUs. MPI is a standard API available to a parallel application that consists of many processes that work on separate processing units (possibly on different nodes of a cluster). Processes interact with each other by exchanging messages. Moreover, a process can include several threads that are allowed to invoke MPI functions provided the MPI implementation supports one of the multithreading levels. This, apart from MPI non-blocking functions, allows overlapping computations and communication. Furthermore, important features of MPI include:

- potentially hardware-optimized collective communication routines,

- parallel I/O allowing several processes to access and modify a file(s),

- dynamic spawning of processes,

- partitioning processes into groups and using various communicators,

- additional functions of certain MPI implementations such as transparent checkpointing in LAM/MPI and BLCR [214].

Development of MPI programs, access and management usually involves the following steps (after logging into the HPC system via SSH):

1. development of source code (e.g., in C/C++ or Fortran),

2. compilation using a compiler with MPI libraries (such as using `mpicc`),

3. submission into a queuing system such as Portable Batch System (PBS), Load Sharing Facility (LSF), or LoadLeveler [141] or running from the console,

4. checking status and browsing results of the application which can be done using the respective commands of queuing systems.

There exist higher-level graphical interfaces that hide details of application submission, checking status and browsing results. Usually these are interfaces built on top of grid middleware able to access several HPC clusters as described in Section 2.1.5. Examples of such interfaces include BeesyCluster, co-developed by the author [69], or MigratingDesktop [12, 136], developed within the CrossGrid project.

MPI programs can be combined with APIs for shared memory programming such as OpenMP [112]. Furthermore, in clusters in which nodes are equipped with GPUs, MPI can be used together with, e.g., NVIDIA CUDA in order to exploit all parallelization levels. When nodes have Intel Xeon Phi coprocessors installed, a single MPI application can be launched in such a way that some processes run on CPU cores and others on cores of coprocessors. There are also other solutions that can help use the available resources in cluster environments. For instance, rCUDA [17, 86] allows sharing GPUs among nodes of a cluster in a transparent way. Many GPUs Package (MGP) [39] through implementation of the OpenCL specification and OpenMP extensions enable a program to use GPUs and CPUs in a cluster. KernelHive [79] allows us to define, optimize and run a parallel application with OpenCL kernels on a cluster with CPUs and GPUs. It is possible to use various optimizers such as minimization of execution time and a bound on power consumption of devices selected for computations.

2.1.2 SOA

Service Oriented Architecture (SOA) [88] assumes that a distributed application can be built form distributed *services*. The service is a software component that can be developed by any provider and made available from a certain location through well-known interfaces and protocols. Such loosely coupled services can be integrated into a complex application that uses various services to accomplish particular tasks.

The popular Web Service technology can be viewed as an implementation of SOA with the following key standards and protocols [185]:

1. SOAP [212]—a protocol that uses XML for representation of requests and responses between the client and the service.

2. WSDL (Web Service Description Language) [213]—a language used to represent a technical description of a Web Service. It contains operations available within the service with specifications of input and output messages, used data types, the location of a service and protocol binding.

3. UDDI (Universal Description, Discovery and Integration) [21]—a standard aimed at development of service registries that could be used by the community. It allows publishing data in a secure way and querying the registry for business entities or services published by them. Providers and services can be described by means of classification systems as well as contain technical characteristics or links to WSDL documents.

The technology features:

- *Self-contained services*—each service is published by its provider and its code is contained within the service and possibly other services it may use. The client does not need any other software other than for encoding a request so that it is accepted by the service and decoding its result,

- *Loosely coupled services*—services are published independently by their providers and are *interoperable* thanks to SOAP.

The standard sequence that allows us to consume a Web Service involves the following steps:

1. discovery of a service that performs a desired business function,

2. fetching the description of the service,

3. building a client and invoking the service.

Figure 2.5 presents a sequence which has been extended compared to the one cited in the literature [185] by inclusion and application of additional standards and protocols to improve the discovery process with semantic search. Semantic computing [196] is gaining much attention in distributed information systems. The steps using these standards are as follows:

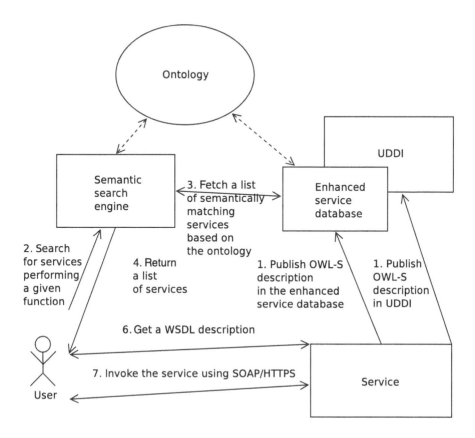

FIGURE 2.5: Publishing, semantic search for services, and service invocation sequence.

1. Publishing information about the service using the OWL-S [82]. OWL-S allows us to specify:

   - Inputs,
   - Outputs,
   - Preconditions,
   - Results.

   Proper concepts/OWL-classes can be used in these fields. Next, the OWL-S description can be mapped into an UDDI description [198].

2. The client sends a request to a UDDI server [185]. In the traditional approach, it is the UDDI server that will search for the service among its records. However, searching in UDDI has been limited to matching keywords with no semantic relation between concepts present in the

search request and the description in the registry. Thus, instead of the UDDI server, the request can be routed to a more advanced semantic engine that will consider an ontology for the search. Namely, for each of the concepts on a particular field of the specification of the request, the engine tries to match and find semantic links to the concepts of the corresponding field of the service description. Paper [198] describes how such semantic connections can be determined using inheritance of concepts.

3. A list of services matching the client's request is returned. The client selects a service and fetches its description in WSDL.

4. The client invokes the service using SOAP, which is transferred using one of many possible protocols, the most common being HTTP and HTTPS.

5. After the service has processed the request, a response wrapped into SOAP is sent back to the client.

The RESTful service [111] solution is an alternative to SOAP-based services. REST (REpresentational State Transfer) [92] distinguishes resources identified by URIs and uses HTTP methods to manipulate the resources. As such, REST can be thought of as a technology allowing realization of Resource Oriented Architecture (ROA) [145]. ROA distinguishes the concept of a resource that is accessible through certain interfaces. Such services can be described by either WSDL 2.0 [151, 213] or Web Application Description Language (WADL) [110]. Technically, REST can also be thought of as an implementation of a (part of) SOA [89].

What is crucial is the possibility of invoking not only one service but to combine outputs of several services and process these further in order to produce output of a given service. This can also be done at more levels of such a service invocation hierarchy. For instance, a travel agency could publish a service for booking a package holiday. After placing a request by invocation of the service, the service would need to find out what other services are available out there to perform more concrete tasks such as:

- booking a flight,

- booking hotels,

- booking additional activities, attractions, events,

- checking weather forecasts for particular places.

Note that there may be several providers offering a particular service, e.g., booking a flight, in which case there will be several alternative services that perform the requested function. Such services can differ in quality metrics such as price, time to book, reservation period, reliability, conditions to cancel, etc.

It can be noted that it may be the travel agency that selects the best flight out of those offered by various airlines or it can further delegate this task by invoking a service of another company that specializes in this very process and does that better for a reasonable fee. Proper selection of such services and execution of such a complex scenario is discussed in depth in Section 2.2.

2.1.3 Multitier Architectures

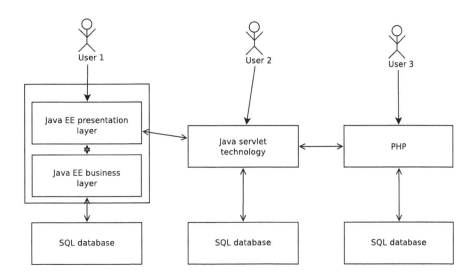

FIGURE 2.6: Multitier architecture.

The multitier approach in distributed system design distinguishes multiple, functionally distinct layers that can be integrated into an application. This has several advantages such as:

- independent development of layers,

- possibility of independent modification,

- possibility of invoking a certain layer from another application.

It is important to note that each layer of such a system exposes its own API to its clients. Typical layers include:

1. The client can be a thin client (Web browser), application running in a sandbox (such as a Java applet), or a desktop application.

2. Server layer that accepts user input, processes it, and sends results back. In some technologies it can be further broken into more layers such as:

- presentation—for verification, parsing input data passed by the client of this layer and presentation of results,

- business—for actual processing of the request and performing the requested function; this layer might contact other layers or other cooperating systems to do its job.

3. Database for storage and retrieval of data. The database engine can aggregate and return only results requested by the client of this layer.

Figure 2.6 depicts three multitier systems with popular technologies and popular APIs exposed by the layers.

2.1.4 Peer-to-Peer, Distributed Software Objects, and Agents

Peer-to-peer computing is a concept of distributed computing in which distributed components cooperate by being able to serve as providers and clients at the same time. In principle, the components are equal and can divide particular tasks of the scenario and management of data among themselves.

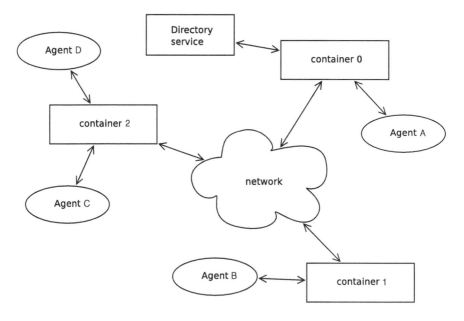

FIGURE 2.7: Distributed agents architecture.

The distributed software objects approach distinguishes distributed objects as components that can interact with each other by invoking exposed methods that operate internally on object attributes. An example of such type of computing is CORBA (Common Object Request Broker Architecture) [194, 115], which allows communication between distributed objects using an Object Request Broker (ORB) [164]. Objects can be implemented in various languages and run on one of many operating systems, in particular using:

- VBOrb [23] for Visual Basic,

- MICO [175] for C++,

- omniORB [13] for C++ and Python,

- ORBit2 [15] for C and Perl, or

- IIOP.NET [8] implementation of IIOP for the .NET environment.

The publish and invocation sequence involves the following steps:

1. Specification of an object interface using the OMG Interface Definition Language (IDL).

2. Translation of the interface to a skeleton on the server side or a stub on the client side. These codes are generated in the language the corresponding side will be implemented.

3. Augmenting the code with actual service implementation on the server side or service invocation on the client side.

4. Creation of objects and storing and passing of Interoperable Object References (IORs) to the client or object registration in the Naming Service (see below).

5. Calling object methods by other objects.

Objects can use services [93] such as the Naming Service, Security Service, Event Service, Persistent Object Service, Life Cycle Service, Concurrency Control Service, Externalization Service, Relationship Service, Transaction Service, Query Service, Licensing Service, Property Service, Time Service, Trading Service, and Collection Service.

Distributed Software Agents [42] can be treated as further extension of distributed objects and be applied on top of the peer-to-peer processing [155]. Particular agents are instances of agent classes and can communicate with each other by sending messages. The following key features of multiagent systems can be distinguished:

- *Autonomous agents*—agents act independently and can perform own decisions,

- *Global intelligence but local views of the agents*—agents can cooperate to achieve a global goal but each agent has its own perception of the part of the environment it can observe. Together agents can acquire and use knowledge that would not have been available for a single party,

- *Self-organization*—cooperating agents can arrange parts of a complex task themselves; on the other hand, agents may need to look for new agents to fetch information, trade, etc., which requires:

  - *negotiation* with other agents,
  - *ontology* as a means of common understanding of used terms for communication between agents that do not know each other.

- *Migration*—agents can migrate throughout a distributed system to gather and use information; this approach might be more efficient than multiple client-server calls; this approach can also cope with partitioning of the system which would make calls from a distant client impossible.

Figure 2.7 presents the architecture of a distributed system which includes containers on which distributed agents work. The agents need a directory service to find and communicate with other agents. An example of such a platform is JADE [42] in which agents have defined behaviors and can communicate with each other by sending ACL messages. Each agent can have its own logic. Two special agents are available: AMS for enabling launching and control of agents and a Directory Facilitator (DF) that allows agents to advertise their services.

It should be noted that agents can be exposed to Web Service clients using the JADE Web Services Integration Gateway (WSIG) [118].

2.1.5   Grid Systems

Grid computing [97, 98, 96] can be best defined as controlled resource sharing. It was proposed as a means of integration of various resources so that there is a uniform way and API for using these resources irrespective of where the resources are, how they differ, or who has provided them. Theoretically, computing resources could be used much like suppliers of electric energy, i.e., connect to the system and use it with the system taking care of which resources are actually used for the request.

Grid computing specifies so-called Virtual Organizations (VOs) that are separate units in different administrative domains. The VOs have certain computing resources at their disposal that can use specific

1. operating systems,

2. storage systems,

3. user authentication and authorization mechanisms,

4. policies of how, when, and if the system can be accessed from outside; the latter can also differ on whether the user belongs to a particular group of users,

5. access protocols and APIs (e.g., WWW, SSH, etc.).

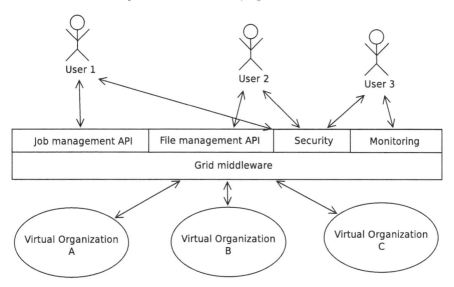

FIGURE 2.8: Interaction between the user and grid middleware and between grid middleware and virtual organizations.

The grid system couples resources of two or more Virtual Organizations into one system. This is done through a particular *grid middleware* that hides all the aforementioned differences among VOs and offers a uniform API to the client of the grid system. The client can use the API and the grid can take care of where the task is executed (Figure 2.8).

Actual implementations of grid systems such as Globus Toolkit [7, 197] offer APIs that allow the following:

1. execution of applications on the grid;

2. management of data on the grid including transfer of files to/from the grid and between particular grid locations; input file staging and output file staging, as well as cleanup after execution of a task;

3. security management including

   • obtaining security credentials to access the grid,

   • right delegation;

4. monitoring resources.

Globus Toolkit is probably the most popular grid middleware. Its version 5 features the following components:

- Grid Resource Allocation Manager (GRAM) [206] for location, submission, canceling and monitoring jobs on the grid by communication with various job schedulers.

- GridFTP [207] protocol for secure, reliable and high-performance data movement in WANs.

- Grid Security Infrastructure (GSI) [208] for authentication, authorization and management of certificates. It uses X.509 certificates and allows delegation through proxy certificates.

- Common Runtime.

The previous Globus Toolkit 4.x version implemented the Open Grid Services Architecture (OGSA) and Web Service Resource Framework (WSRF) [53, 197] and additionally provided a Monitoring and Discovery System (MDS).

Other available grid middleware include UNICORE [22, 35, 201], and gLite [87, 138].

On top of grid middleware, several high-level interfaces and systems have been developed that make running applications, management of data, and information services easier to use. These do allow us to run specialized parallel applications in the grid environment, the former often tailored to the needs of specific domain users. Examples include, among others, CrossGrid [105], CLUSTERIX [219, 131, 134], EuroGrid [139], and more recently PL-Grid [51, 43, 130] in Poland. These do require additional functionality compared to stateless Web Services as the client-server interaction often requires a context to be retained between successive calls for the identification of the client to return, e.g., results of previous requests. Furthermore, the client must be prevented from unauthorized access to resources they should not have access to. Accounting, secure data transmission and resource discovery are crucial for distributed computing between virtual organizations.

2.1.6   Volunteer Computing

Volunteer computing allows distributed execution of a project by a large group of geographically distributed user volunteers. Similar to grid computing, it is about integration of resources (especially computational) in a controlled way. However, there are also key differences. Table 2.1 lists both similarities as well as different solutions adopted in these approaches.

The simplest architecture of a distributed volunteer-based system can contain one project server with multiple distributed volunteers connected to it. The client computers perform the following steps in parallel:

1. Define usage details, i.e., CPU percentage/cores, RAM, and disk, that can be used by the system.

2. Download computational code from the project server.

TABLE 2.1: Grid vs volunteer computing.

Feature	Grid Computing	Volunteer Computing
Integration of resources in controlled way	+	+
Participating parties	Virtual Organizations (VOs)	End users
Large-scale distributed processing	+	+
Suitable only for compute-intensive applications without dense communication	+	+
Uniform job submission and file management API for clients who want to use resources	+	—
Security and reliability of computations	Trusted VOs	End users cannot be trusted; need for replication
Social undertaking	—	+
Safe for the client	+ (Node code executed on the client side)	-/+ Needs setting up a sandbox such as a separate UNIX account for real safety

3. in a loop:

   (a) download a packet(s) of input data,

   (b) process the data,

   (c) report results and effort in terms of CPU and memory used over the processing period.

The aforementioned scheme is typical of volunteer-based systems such as BOINC [30]. The BOINC system allows running several independent projects that attract volunteers to donate processing power, memory and storage to process data packets. Management of such a project, keeping the volunteers active and attracting new ones becomes a socially oriented initiative. The BOINC server may need to send two or more packets of data to various clients to make sure that returned results are correct. This may allow us to assess reliability of particular clients over time and rely on single processing requests. Similarly, assessment of the reported effort is based on the minimum out of reported efforts from two volunteers.

Comcute [6, 37, 75] is a system for volunteer computing in which the server layer is extended compared to BOINC for purposes of reliability and security of computations. Contrary to BOINC, computations on the client side are

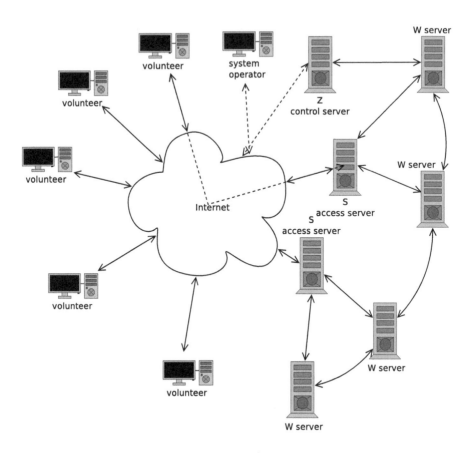

FIGURE 2.9: Volunteer computing architecture.

executed within a web browser, thus relieving the client from any installation
of any application. Figure 2.9 presents the architecture of the system with the
following layers:

1. Servers (W) for management of computations; within these servers a
   group of servers is elected for handling the task in a collective manner
   by distribution of data packets among S servers connected to the W
   layer.

2. Distribution servers (S) to which volunteers connect in order to:

   (a) Submit the capabilities of the client system, i.e., the computing
       technologies it supports, e.g., JavaScript, Java, Flash, Silverlight.

(b) Obtain computational code; the server can use the information provided in the previous step to provide the code in the most efficient
technology.

(c) Successively fetch input data and return results.

3. Access server (Z) that exposes the access API and a Web interface as an
indirect way to access W servers; for security reasons it is not possible
to connect to W servers directly.

In Comcute it is possible to define the logical structure of W servers, connections between S and W servers, redundancy used when sending requests to
particular volunteers, and the number of W servers in charge of a particular
project. From the API point of view, the programmer needs to provide code
for a data partitioner, computations and data merger.

2.1.7 Cloud Computing

Cloud computing [29] has become widespread and popular in recent years.
The client is offered various kinds of services that are made available from *the
cloud* and managed by providers. Be it infrastructure, particular software, or
a complete hardware, operating system, and a software platform, all options
are marketed as being:

1. *cheaper* than purchasing the necessary equipment and/or software li-
censes by the client, necessary maintenance, administration, introducing
software updates;

2. *scalable* because the client can often scale the configuration not only up
but also down depending on current needs;

3. *convenient and easy to use* as the client does not need to focus on hardware and software management if the core business is not related to
computer science; in this scenario the client does not deal with any
hardware failures, software incompatibilities, etc.;

4. *accessible* as the cloud can be accessed from practically anywhere using
any device assuming the Internet is available.

On the other hand, the following should be noted:

1. *The data is stored in the cloud* and is handled by the provider and the
client does not always know where and how the data is stored. Depending
on where the provider offers their services and where the data is located,
various laws may govern; consequently the legal aspects might not be
clear for the client unless specified clearly in the agreement.

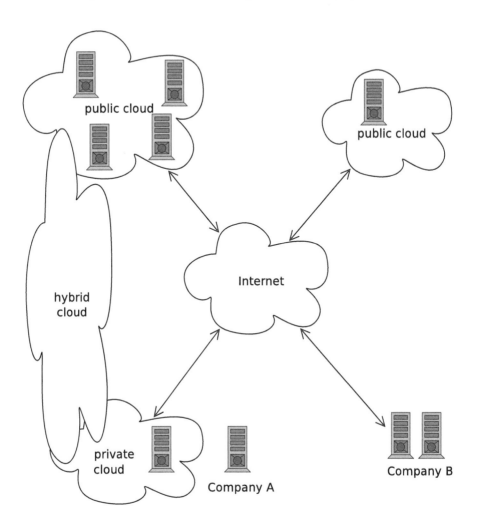

FIGURE 2.10: Cloud computing with private, public, and hybrid clouds.

2. *Risk of vendor lock-in*: If the vendor started changing conditions considerably, the client may find it difficult to migrate to another cloud provider if too invested; it might be time consuming and costly (however some providers use compatible solutions).

Based on available solutions and particular offers including parameters and prices as well as specific functions, a client may create a configuration possibly using services from many providers [189].

Alternatively, institutions my want to set up private clouds to mitigate these issues. The following types of clouds can be distinguished in this respect:

- *Public cloud*, in which case an external provider offers access to a cloud or clouds managed by them.

- *Private cloud*, which is maintained by the given institution and offered as a service within this institution. If the latter is large, it may turn out to be an efficient way of resource management for various geographically distributed subsidiaries. At the same time, the cloud is managed and controlled by the same institution. However, it is the institution that needs to purchase and maintain both hardware and software in this case.

- *Hybrid cloud*, in which particular services may be provided from such a location as to optimize factors required by the client; as an example:

  1. An application for which high security is required, may be provided from a private cloud while infrastructure for time-consuming but not that critical computations may be used from a public cloud.

  2. An enterprise can continue to store sensitive data of its current clients and contracts in a private cloud and offload less critical archive data to a public cloud.

  3. An enterprise interested in particular software typically offered by a public cloud may use a private cloud for this software due to security reasons.

The following sections present types of cloud services offered on the market. Each of the services is provided based on a contract with a certain duration and certain QoS terms, including the price.

2.1.7.1   Infrastructure as a Service

Infrastructure as a Service (IaaS) is a type of a cloud service that offers clients the requested resources in a particular configuration. The client can request particular parameters with respect to the following:

1. Number of requested nodes/computers

2. RAM size

3. Storage/disk space

4. CPU power

5. Operating system, e.g., Microsoft Windows, Linux etc.

6. Outgoing and incoming transfer per month

7. QoS parameters such as requested availability, e.g., 99.9%

What is important is that the client can usually scale up and down the requested configuration based on current needs and pay only for the configuration currently needed. Examples of particular cloud-based compute-type solutions include Amazon Elastic Compute Cloud (Amazon EC2) [156] and Google Compute Engine service [171]. Another solution is Eucalyptus (with support for Amazon EC2 and Amazon S3 interfaces) which allows creation of private and hybrid clouds [90]. OpenStack [14, 179] allows you to manage compute, storage, and networking resources and is a platform for private and public clouds. It also provides an API compatible with Amazon EC2.

2.1.7.2   Software as a Service

Software as a Service (SaaS) is a type of cloud service that offers clients access to particular software. The software is maintained by the cloud provider who performs all necessary updates while the client can use the service from any location. Most often, *flat payment* for the services in the given configuration is assumed and for the duration of the contract.

2.1.7.3   Platform as a Service

Platform as a Service (PaaS) is often referred to as exposing a complete hardware and software platform as a service to clients. For instance, a particular complete programming environment with a requested operating system may be offered on an appropriate hardware system to programmers with a team work environment. Examples include Red Hat's OpenShift [174] and Aneka [135, 170].

2.1.7.4   Cloud vs. Grid Computing

There are a few important differences when compared to grid systems [104]:

1. Business goals: While grid systems emerged as distributed systems for integration of high-performance resources of mainly institutions, cloud computing's targets are both businesses and end users who outsource hardware, software, or hardware and software platforms.

2. API: Grid systems use grid middleware that can use queuing systems to access particular computational resources. Grid middleware expose job management, file management, monitoring and security APIs as explained in Section 2.1.5. Cloud systems use specific APIs or interfaces such as Amazon EC2 and Amazon S3.

3. Organizational: In grids: cooperation including allowing access to use resources, authentication and authorization of users is agreed between Virtual Organizations. In cloud computing, providers generally attract clients independently.

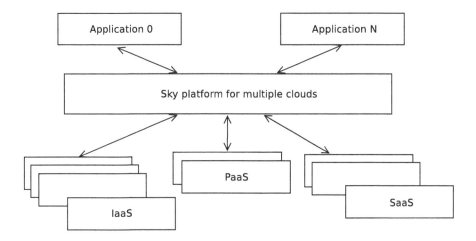

FIGURE 2.11: Sky computing on top of multiple clouds.

2.1.8 Sky Computing

Obviously, there are many IaaS, SaaS and PaaS providers. This means that, similarly to services in SOA, a client can select the best cloud providers based on required QoS parameters. For instance, Cloudorado [5] allows users to find the best IaaS offers based on the client's requirements.

Sky computing [127] can be seen as an extension and generalization of cloud computing to overcome the limitations of the latter. Namely, the client uses an infrastructure that connects to multiple clouds to use required services as shown in Figure 2.11. Thus sky computing leads to creation of a middleware on top of cloud providers. The goal of this middleware is similar to the functions of the grid middleware. Similarly, issues such as the following ones appear:

1. lack of a uniform interface to various clouds (some use the same interface as explained in Section 2.1.7) as well as security management,

2. difficult communication and synchronization between clouds; there are network restrictions for particular clouds related to configuration of VMs, IPs, high numbers of hops between machines [95],

3. difficult data management across multiple clouds at the same time.

Compared to clouds, sky computing offers the following benefits:

1. Independence from a single cloud provider: In case one cloud changed its conditions considerably, the client can simply select another cloud

provider without much effort; this is obviously possible if the APIs for the clouds are the same or similar.

2. Promotes increased competition from other cloud providers that can be used by a particular client. Sky computing mitigates the problem of *vendor lock-in* in this way. This is beneficial for the client and the provider might be well aware of other options for the client.

3. Ability to offer a set of services on top of clouds that would not be available from a single cloud provider

4. Ability to use various clouds from possibly various providers for particular parts of a complex scenario.

2.1.9   Mobile Computing

The recent years have marked fast growth of the mobile device market. The following important software areas for mobile computing could be distinguished since mobile devices became popular recently:

1. personal information management, scheduler, office applications,

2. client to WWW servers and Web Services,

3. location-based services based on the position of the mobile device,

4. sensors used in smart homes/world such as GPS, acceleration, light, temperature, etc.

The latter is also related to *ubiquitous computing* in which the distributed system is filled in with sensors and proper services are invoked based on the current context. For instance, when the owner of an apartment approaches, the door is unlocked and music is turned on based on the current mood of the owner, temperature, humidity and the time of day. Mobile devices with a wide array of sensors may be used for this purpose as well.

Naturally, mobile devices can communicate with various services around and even far away from them. Two types of services can be distinguished:

1. Those offered by other mobile devices using, e.g., Bluetooth.

2. Services offered from stationary servers, e.g., Web Services, servers (e.g., mail) allowing socket connections, POP, asynchronous notification using, e.g., Google Cloud Messaging (GCM) [114], etc. This can allow integration of mobile devices with weather forecasts, HPC systems for processing of digital media content from cameras, etc.

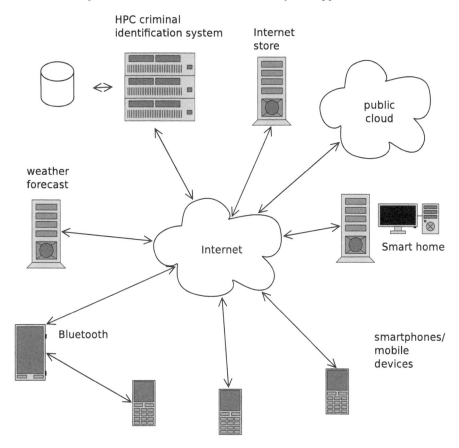

FIGURE 2.12: Mobile computing.

2.2 Complex Distributed Scenarios as Workflow Applications

Today's distributed systems are all about service integration as this leads to added value and possibilities compared to individual services. *Workflow management systems* have been introduced to model, optimize, execute and manage complex scenarios in information systems. In general, a complex scenario is represented by a *workflow application* or a *workflow* which can be modeled as a graph in which nodes correspond to tasks which are parts of a complex process while the edges denote control and corresponding data flows. This approach applies, with slight differences, to complex tasks in the following areas:

1. Business processing in which tasks can be performed by various services offered on the market by various providers on different terms. Business workflows are more about control flow rather than processing large amounts of data; various control mechanisms should be predicted including exceptions.

2. Scientific computing in which workflow tasks are usually to process large amounts data; parallel paths of the workflow can be used in order to parallelize processing; control flow is rather simple.

3. Ubiquitous computing in which the graph depicts sequences of invocations of services based on the context, i.e., a set of conditions that initiates a given service or services.

Workflow applications can be characterized in terms of many features. Typical aspects applicable to many contexts are discussed in subsequent sections.

For applications run on certain system types, thorough taxonomies have been proposed, e.g., for grid workflow applications in [216, 222].

2.2.1   Workflow Structure

Depending on the particular field of study, various constructs are allowed in workflows which determine possible control and data flow as well as optimization and execution. Typical constructs may include:

1. *sequence (S)*—tasks connected with a directed edge are executed in the specified order,

2. *choice (C)*—one of alternative parallel paths can be executed,

3. *parallel execution (P)*—two or more parallel paths may be executed at once,

4. *iteration (I)*—a part of a workflow may be executed many times.

The most popular formulation of a workflow is a directed acyclic graph (DAG) which includes constructs S, C and P.

2.2.2   Abstract vs. Concrete Workflows

Given a particular workflow structure, there are two ways that executable components can be assigned to particular workflow nodes. This determines the type of workflow:

- *Concrete*: In this case it is specified a priori what executable code is assigned to particular workflow tasks. This might be given in various forms:

1. for each workflow node an application (executable) is assigned that executes on a particular resource (computer),

2. an external concrete service is assigned to a workflow node.

- *Abstract*: In this case the workflow definition for each workflow node contains specification of what task the node should do. This can include:

  1. a *functional* description of what the particular workflow node should perform; one of many possible executables or services can be selected to accomplish this task,

  2. *non-functional (QoS)* requirements imposed on the workflow task, e.g., that it must finish within a specified time frame or must not be more expensive than a given threshold.

2.2.3 Data Management

Data management is very important for workflow applications oriented on both:

1. data flow, where options for partitioning of large amounts of data for parallel processing are crucial,

2. control flow, where, e.g., making a choice based on input data is made.

The literature discusses several constructs and approaches: various ways of data integration coming from various inputs [103], data representation such as files, or data streams [60, 106].

It should be noted that workflow tasks may be assigned particular data and correspondingly data sizes in advance. In such cases the problem is usually to determine which service should perform the given task such that a function of quality metrics is optimized.

Many other problems could be considered as well. For instance, if input data of a given size needs to be partitioned among workflow paths for parallel execution in such a way that capacity constraints need to be met and flow conservation equations need to hold as well as the cost of processing is to be minimized, we would consider the minimum cost flow problem [203].

2.2.4 Workflow Modeling for Scientific and Business Computing

The most popular model of a workflow application used in scientific and business applications is an abstract DAG workflow. This allows flexibility in the definition of a complex task, allows selection of appropriate services, and stating a practical optimization problem for subsequent execution. The process involves a few, steps:

1. Workflow definition

2. Finding services capable of executing workflow tasks

3. Selection of services for particular workflow tasks and scheduling service execution at particular moments in time

4. Execution

5. Monitoring the status and fetching output results

Formally, an abstract DAG-based workflow considered in the literature is represented as a directed acyclic graph $G(T, E)$ where T is a set of tasks (graph nodes) and E a set of directed edges connecting selected pairs of tasks. For each task t_i there is a set S_i of services s_{ij}s each of which is capable of executing task t_i. The notation corresponding to this model is as follows:

- t_i—task i (represented by a node in the workflow graph) of the workflow application $1 \leq i \leq |T|$

- $S_i = \{s_{i1} \ldots s_{i|S_i|}\}$—a set of alternative services each of which can perform functions required by task t_i; only *one* service must be selected to execute task t_i $1 \leq i \leq |T|$

- $c_{ij} \in R$—the cost of processing a unit of data by service s_{ij} $1 \leq i \leq |T|$ $1 \leq j \leq |S_i|$

- $d_i = d_i^{\text{in}} \in R$—size of data processed by task t_i $1 \leq i \leq |T|$

- $t_{ij}^{\text{exec}} = f_{ij}^{\text{exec}}(d_i^{\text{in}})$—execution time of service s_{ij}

- $t_i^{\text{st}} \in R$—starting time for service s_{ij} selected to execute t_i $1 \leq i \leq |T|$

- $t_{\text{workflow}} \in R$—wall time for the workflow i.e., the time when the last service finishes processing the last data

- $B \in R$—budget available for the execution of a workflow

Examples of practical workflows for three application areas are as follows:

- *Business*—a workflow application that can be regarded as a template for development and distribution of products out of components available on the market; generalization of a use case considered in [66] and extended with dependencies between services (Figure 2.13):

  1. The first stage distinguishes parallel purchases of components from the market; for each one there may be several sale services available with different QoS terms such as delivery time and cost.

  2. The product is integrated out of the collected components with the know-how of the workflow owner (company).

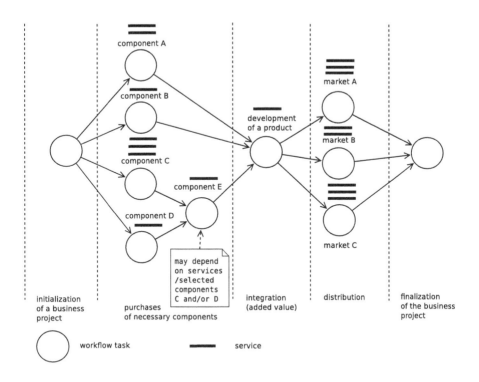

FIGURE 2.13: Business workflow application example.

3. The product is distributed on several, potentially geographically distributed markets; for each market there may be several distributors the product developer can choose from.

- *Scientific*—a workflow for parallel processing of input data shown in Figure 2.14 that can be used for a variety of applications such as parallel processing of aerial images, signals from space, recognition of illnesses based on patient's images and a database of templates, voice recognition, detection of unwanted events [74]. The author distinguished a generalized template with several stages that fit and can be adjusted to all these applications:

  1. Data acquisition—input data can be gathered in parallel from various sources such as cameras, audio input devices, etc. it can be noted that there can be both concrete tasks for which there already known data acquisition services (such as a particular camera) or abstract tasks; for the latter, there can be various services that provide services, e.g., services from various market analysis providers.

2. Data set preparation—acquired data is arranged into data sets for parallel processing; note that both various data sets can be processed in parallel as well as each individual data set may be parallelized in the next stage.

3. Parallel processing of particular data sets. It should be noted that in this workflow formulation it is possible to process various data sets in parallel; in particular, acquisition of another data set (such as from source C) does not synchronize with processing of data from source A; parallel processing has been arranged into parallel workflow paths which are realized by particular services installed on supposedly various resources.

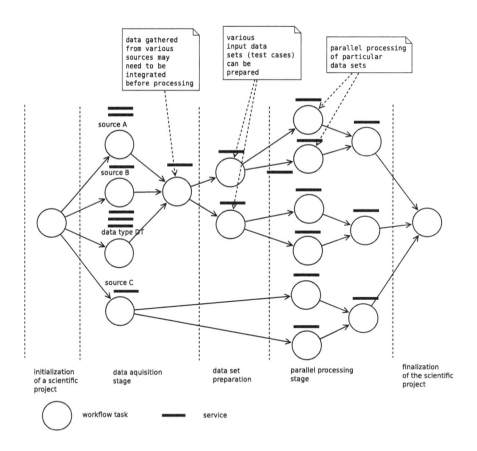

FIGURE 2.14: Scientific workflow application example.

2.2.5   Workflow Scheduling

For abstract workflows that contain tasks with two or more services assigned to them, a workflow scheduling problem needs to be solved. Given the aforementioned dependency constraints defined by the graph, using the introduced notation, the workflow scheduling problem can be stated as follows:

$$\forall_i \text{ find } t_i \rightarrow (s_{i\ sel(i)}, t_i^{\text{st}}) \tag{2.1}$$

where $s_{i\ sel(i)}$ denotes the service selected for execution of t_i starting at time t_i^{st} such that

$$\forall_{i,k:(t_i,t_k)\in E} \ t_k^{\text{st}} \geq t_i^{\text{st}} + t_{i\ sel(i)}^{\text{exec}} \tag{2.2}$$

with constraints on the workflow execution time and cost. The workflow execution time (not considering scheduling time) is the end time of the latest task with no successor:

$$t_{\text{workflow}} = max_{i:\nexists k(t_i,t_k)\in E}\{t_i^{\text{st}} + t_{i\ sel(i)}^{\text{exec}}\}. \tag{2.3}$$

Typically, considered alternative optimization goals include one of the following:

- *MIN_T_C_BOUND—minimization of the workflow execution time with a bound B (budget) on the total cost of selected services:*

$$min\ t_{\text{workflow}}$$
$$\sum_i c_{i\ sel(i)} d_i \leq B. \tag{2.4}$$

- *MIN_C_T_BOUND—minimization of the total cost spent with an upper bound on the workflow execution time T:*

$$min \sum_i c_{i\ sel(i)} d_i$$
$$t_{\text{workflow}} \leq T. \tag{2.5}$$

- *MIN_H_C_T—minimization of a function of cost and execution time (assuming certain units):*

$$min\ h(\sum_i c_{i\ sel(i)} d_i, t_{\text{workflow}}). \tag{2.6}$$

Formulation 2.6 might be useful if we consider a financial equivalent of time which is a very practical approach applicable to, e.g., transportation [152]. Formulations such as 2.4 and 2.5 are NP-hard problems [225] which means that optimal solutions can be found in a reasonable time frame only for problems of small size. For larger problems that show up in real-life situations, efficient heuristic algorithms need to be adopted. Selected algorithms for solving these problems are discussed in Section 4.2.

2.2.6   Static vs. Dynamic Scheduling

It should be noted that although the workflow scheduling problem considers a set of services for optimization, availability of these services might change. In general, in real environments, the following types of events might occur and impact scheduling results:

1. Some services that were previously chosen for execution are no longer available. This might be due to failure of a network, the server on which the service is installed or possibly even maintenance.

2. Changes in service parameters unless a contract was put in place that would guarantee expected values for QoS metrics such as execution time, price, etc.

3. New services showing up on the market that would be capable of executing workflow tasks on better terms.

In cases 1 or 2, rescheduling of the workflow has to be performed in order to meet the optimization goal with possibly alternative services. In case 3, the cost of rescheduling should be assessed as compared to potential gains from using services with potentially better parameters.

It should be noted that there can be many more parameters/quality metrics assigned to services, apart from cost c_{ij} and time t_{ij} such as dependability, reliability, security, being up-to-date, etc. These are discussed in Section 2.3.3. All of these can also be incorporated into either the aforementioned or dedicated constraints.

2.2.7   Workflow Management Systems

There are several workflow management systems available which can be characterized in particular by [222]:

- *Target application type*—business, scientific, ubiquitous computing

- *Underlying type of services*—e.g., one of the following: Web Services, grid or cloud services

- *Manual composition of tasks into workflows or automatic based on existing rules and facts*

- *Optimization* of a single criterion or multiple criteria at the same time

- *Scheduling*—using local or global knowledge

- *Various ways of monitoring and learning* about services and providers, etc.

Exemplary workflow management systems that can be distinguished based on the type of target systems include:

- *Grid*—usually built for grid systems on top of grid middleware such as Globus Toolkit, Gridbus, etc. or services. Examples of such systems include ASKALON [215], Taverna [18], Pegasus [83, 84], Triana [20, 150], Kepler [11, 146], Conveyor [142], and METEOR-S [137], which allows adding semantics to web service composition with quality of service [27, 176].

- *Cloud*—recently, cloud systems have become attractive platforms for both business and scientific services [45, 63] with emphasis on the payment-per-use policy for cloud systems compared to grid systems [116, 211] and possibility of launching various configurations easily. Another related advantage is scaling resources as needed, which can be useful for scientific workflow applications with a varying level of parallelism such as different numbers of parallel paths in various stages. Dynamic scaling is possible in the Aneka Cloud [170]. Cluster-based systems offer better communication performance at the cost of flexibility [123]. Examples of systems include:

  - Amazon Simple Workflow [40]—allows definition and management of business workflows on clouds and/or traditional on-site systems.

  - Tavaxy [26]—a system that integrates Taverna and Galaxy for launching a part or a whole workflow in a cloud.

  - A framework and workflow engine for execution of workflows on many clouds [101].

- *Ubiquitous computing*—services are invoked in the given context. FollowMe [140] is an example of a platform for this type of computing.

Similarly, various languages and standards for representation of workflow applications are used in various contexts [104], for example:

- *Grid*—Petri-Nets (e.g., Triana) [150], Abstract Grid Workflow Language (AGWL) [91] used in ASKALON

- *Business*—Business Process Modelling Notation (BPMN) [165] with XML Process Definition Language (XPDL), Web Services Business Process Execution Language (WS-BPEL) [205], OWL-WS [41] (NextGrid). Orchestra [16] is a solution for complex business processes with WS-BPEL 2.0 support. BPELPower [221] was designed and implemented in

Java as a solution that can run both typical BPEL and geospatial workflows with handling Geography Markup Language (GML) and geospatial Web services. ApacheODE [2] is able to execute processes expressed using WS-BPEL 2.0 and the legacy BPEL4WS 1.1. Activiti [1] is a platform with a BPMN 2 process engine for Java. jBPM [10, 148] is a Business Process Management (BPM) Suite with a Java workflow engine for execution using BPMN 2.0. Execution using BPEL is considered in various contexts such as grids [147] or mobile devices [109].

- *Ubiquitous computing*—Compact Process Definition Language (CPDL) in FollowMe [140].

SHIWA (SHaring Interoperable Workflows for large-scale scientific simulations on Available DCIs) [24] allows you to store various workflows in a repository and run these on various Distributed Computing Infrastructures (DCIs). What is novel is that the approach enables you to define higher-level meta-workflows that consist of workflows run on various workflow management systems. This greatly increases interoperability among workflow-based solutions. Furthermore, Askalon was extended for use in grid-cloud environments [167] with an Amazon EC2-compliant interface. Launching a workflow on various systems including clouds, grids and cluster-type systems is shown in [170] with support for Aneka, PBS, and Globus.

2.3   Challenges and Proposed Solutions

2.3.1   Integration of Systems Implementing Various Software Architectures

As described in Section 2.1, various software architectures of distributed systems were designed for particular types of processing and applications. Consequently, integration of various components within a single software architecture and its actual implementation is straightforward. For example:

1. HPC: MPI allows parallel processing and synchronization of processes running on various nodes or processors.

2. HPC: NVIDIA CUDA and OpenCL allow parallel processing using multiple threads running on many Streaming Multiprocessors (SM). The same technologies allow parallel usage of multiple GPUs at the same time.

3. SOA: By design, a Web Service can consume and integrate results of other Web Services. Web Services are loosely coupled and can be offered by various providers, from any place. Cooperation is possible using the well-established SOAP-based or RESTful services.

4. Multitier Architectures: Within Java Enterprise Edition [120], components of the presentation layer such as servlets and JSPs cooperate easily with Enterprise Java Beans in the business layer. Furthermore, components can interact across application servers thanks to clustering within this technology [169].

5. Grid systems: By design, systems and software of distributed Virtual Organizations (VOs) is coupled together and made available using a uniform API and interface that is called *grid middleware*. As an example, Globus Toolkit provides a uniform API for management of jobs (tasks) on the grid such as clusters located in various VOs through GRAM, file management through GridFTP, and security using GSI.

6. Cloud computing: Various resources offered by a cloud provider can be used together seamlessly and in a way not visible to the client, e.g., in Google Apps service.

7. Mobile computing: Well-established standards such as Bluetooth, and HTTP serve as common communication protocols for various mobile devices. Nevertheless, even though Bluetooth is a well-defined standard, communication glitches including disconnecting, and problems when establishing communication are common when using devices such as mobile phones and GPS navigation devices from various manufacturers. Other standards such as Web Services, and communication over sockets can be used as well-provided there is an implementation for the devices in question.

Problem: Integration of applications or components running within software systems implementing *different* software architectures is not always possible out-of-the-box. There do exist some approaches in various contexts. For instance, *sky computing* [127] aims at integration of various clouds from various providers. In an enterprise environment, Enterprise Service Bus (ESB) allows integration and communication between various applications and services using various communication protocols and formats [182, 202]. Various topologies are possible for integration of, e.g., separate subsidiaries: a unified or separate ESBs [143]. A BPEL engine that integrates services can be either independent from an ESB and can call services from various ESBs or operate within an ESB [143]. As [182] suggests, extension of an ESB implementation with new protocols is not easy as it demands a new port type. A concept of the universal port is proposed along with protocol and format detectors and processors for use with an ESB. Furthermore, it is always possible to use certain technologies such as Web Services in order to create a middleware hiding differences between certain types of systems standing behind it.

However, at the level of the aforementioned software systems implementing particular software architectures there is no solution that would address all aspects of uniform integration. Such a solution requires integration at the level of protocols, and data formats, but also authentication, authorization, potential queuing systems and potentially complex APIs exposed by those software systems.

Rationale: For many modern applications, integration of various distributed systems is desirable. For instance:

1. business applications using services from various companies, entities including government and local authorities,

2. scientific computing for gathering data from, e.g., distant radio telescopes from all over the world and processing in parallel on various clusters,

3. ubiquitous computing when local services are discovered and used dynamically in a workflow application based on the physical location of the device.

As an example, the scientific workflow shown in Figure 2.14 may use services for parallel processing of data, whether on an HPC system with the PBS queuing system, available as a Web Service from a distant server, available through Globus Toolkit, or processed in parallel by many volunteers attached to BOINC. From the perspective of the workflow modeling, it does not affect the structure or definition of the workflow in any way. Such mapping of software implemented using various architectures could prove useful.

Proposed solution: The author proposes to use a generalized and uniform concept of a *service* for *any* software components implementing these software architectures to be able to integrate actual resulting services into complex scenarios modeled by workflow applications. In fact, the concept of a service is already present in, e.g.,:

1. SOA—as a basic software component and implementation through the Web Service [88]

2. Grid computing—the concept of a *grid service*

3. Cloud computing—the concept of SaaS

Some other architectures and actual implementations require publishing of selected legacy software components as services. The actual solution is presented in Sections 3.1, 3.2, and 3.3.

2.3.2 Integration of Services for Various Target Application Types

Essentially, software services can be divided into categories in terms of target application types:

1. Business—when the functional goal of the service is customer oriented with QoS metrics such as performance, cost, reliability, conformance, etc. are very important for the client.

2. Scientific— when the primary goal of the service is to perform computations, data analysis, predictions, etc., that do not need to be directly applicable or purchased by an average consumer. Usually such services are performance oriented although the cost and power consumption are gaining attention as shown in Section 2.1.1.

3. Ubiquitous— services oriented on consumers but usually responding to frequent requests with fast replies and responding in the given *context*.

Problem: Provide a concept of a software component that would contain a uniform description of the purpose, target environment, and handling a request and response as well as offer a uniform API for client systems.

Rationale: Many modern interdisciplinary problems require services from various fields of study. For instance, design of a modern aircraft would require simulations related to strength, durability, reliability of various components that would need to be performed on HPC resources, most likely in different, geographically distributed centers, purchase and service cost analysis, marketing services and many others. This will certainly involve business, and scientific services performed both by software and human specialists.

Proposed solution: The author proposes a uniform description of the *service* as a component suitable for all these target uses. The description would contain all details relevant for all these target applications. The actual solution is presented in Sections 3.1, 3.2, and 3.3.

2.3.3 Dynamic QoS Monitoring and Evaluation of Distributed Software Services

Uniform description of services targeting various applications and internally implemented using various software architectures does not make these practical to use until one can reliably assess their qualities. What is more, it can be seen clearly in today's distributed software market that static assessment of the service QoS is not enough. Table 2.2 lists QoS metrics for which precise, reliable, and up-to-date assessment for particular service types is crucial. Thus, the following can be stated:

TABLE 2.2: Important QoS metrics for particular types of services.

Service Type	QoS Metric	Notes
	Performance	Usually HPC is used to shorten the execution time
HPC	Dependability	One expects to trust the returned results
	cost	Usually needs to be below the predefined threshold

SOA	Availability and reliability of a service	What the customer expects from the service
	security	All communication between the client and the service should be encrypted and data processed and/or stored in a secure way
Multi-tier Systems	Ease of use	Clear and friendly API
	Availability	From the customer's point of view
Grid	Ease of use	Availability is supported by the grid middleware that would make use of available services on attached Virtual Organizations
	Security	Crucial when cooperating with other organizations
Volunteer	Performance	The main reason to search for others' computing resources
	Reliability	Results need verification against hardware/software errors and potential harmful users' intentions
Cloud Computing	Security	Crucial for the client because the data is managed by a third party
	Availability of alternatives	To avoid vendor lock-in
	cost	The client wants to minimize it
	reliability	The customer expects the service (IaaS, SaaS, PaaS) to be working at any time within Service Level Agreement (SLA)
	Performance	The client wants to maximize it within the cost constraint
Mobile Systems	Availability	The customer expects the service to be mostly accessible
	Cost	May be a decisive factor for choosing the service as there may be many alternatives
	Location awareness	May help optimize the result of the service
	being up-to-date	Must be based on up-to-date data, e.g., latest maps, information about Points of Interest (POI), etc.

Problem 1: Perform periodic, reliable monitoring and analysis of QoS metrics important for particular types of services. If needed, prepare forecasts for QoS values in the future that result from the past values.

Problem 2: Perform periodic, up-to-date ranking of services (possibly implemented on various architectures) that perform the given function considering past history of quality evaluation.

Rationale: Most of the aforementioned QoS metrics for the distinguished types of services change in time. Thus, it is necessary to:

1. discover new services dynamically,

2. monitor QoS of services dynamically, which may hint that some previously preferred services are no longer available.

Furthermore, it is necessary to evaluate services that are capable of performing the function requested by the user. This leads to ranking services. Depending on particular needs, various algorithms may be necessary also considering the history of QoS evaluations for the given service.

Moreover, from a global point of view, the ranking scheme may need to assure that no single provider dominates the market with their services to avoid the vendor lock-in problem in cloud computing.

Proposed solution: The suggested uniform definition of the service is extended with the procedure that describes how:

1. particular QoS metrics of the service will be measured, including dynamic measurements and application of digital filters;

2. aggregation of QoS measurements into a ranking of services in the given category.

The actual solution is presented in Section 4.1.

2.3.4 Dynamic Data Management with Storage Constraints in a Distributed System

Today, data handling becomes one of the key concerns. As computing power is becoming available almost everywhere (clusters, servers, personal computers, multicore tablets and smartphones), data management, storage constraints and moving data toward compute devices to optimize QoS criteria, especially in complex workflow applications becomes crucial. This holds true not only for particular types of systems but especially for integration of the aforementioned types of systems including sky computing.

Problem 1: Consider storage constraints in execution of complex scenarios that incorporate distributed services. Consider data communication costs in optimization of execution of the scenarios, especially location of management units and data caching.

Problem 2: Consider different ways of handling data: *synchronized* or *streaming* in various stages of a complex distributed scenario.

Rationale: Although larger and larger storage spaces are available and the price per storage is reasonably low, it is not free. It may be especially important for scenarios which handle large amounts of data, e.g., processing sales, stock data in data centers, and multimedia data out of cameras installed in malls, city centers, and along roads. Especially the latter would need transfers of large amounts of data from distributed sources, staging to computational resources, caching, parallel processing, and storage of data from a certain period from the past. Big data processing is one of the directions on which the current research in information systems is focused [153, 33].

Proposed solution: An integrated solution that distinguishes:

1. storage constraints of the resources on which services are installed,

2. storage constraints of the system that executes various services of a complex scenario and transfers data from one resource to another, and

3. caching of intermediate data transferred by the workflow management system in a dedicated cache storage.

The actual solution is presented in Chapter 5.

2.3.5   Dynamic Optimization of Service-Based Workflow Applications with Data Management in Distributed Heterogeneous Environments

Problem: Define a quality model, a dynamic optimization problem and solutions for complex scenarios that would correspond to real-life, often interdisciplinary, complex tasks that couple services from various domains.

Rationale: Complex interdisciplinary tasks often require services from various fields of study. These services differ in terms of the

1. application domain,

2. implementation platform,

3. important QoS metrics, and

4. evaluation criteria of particular services.

Proposed solution: A workflow scheduling model that integrates all of the following:

1. a model that considers a uniform concept of a service as indicated in Section 2.3.1 with a description independent from the application target as indicated in Section 2.3.2,

2. runtime monitoring and evaluation of services as suggested in Section 2.3.3,

3. definition of the workflow scheduling problem considering dynamic evaluation of services and rescheduling along with management of data processed by services as hinted in Section 2.3.4.

The detailed solution is presented in Section 3.4 with algorithms presented in Chapter 4.

Chapter 3

A Concept of Dynamic Quality Management for Service–Based Workflows in Distributed Environments

3.1   A Generalized Quality Model of a Service

Proposal of a generalized model of a service needs to be preceded by analysis of the types of distributed software used in today's distributed systems. Such characteristics were described in Section 2.1 for the following:

1. HPC systems,

2. software services in SOA,

3. components in multitier architectures,

4. distributed software objects and autonomous agents,

5. grid services,

6. software working in the volunteer computing concept,

7. cloud services and services in federations of clouds, and

8. mobile software.

In order to incorporate any type of service into a workflow application or adapt legacy software to services, the author proposes a uniform conceptual model of a service shown in Figures 3.1 and 3.2. Namely, the analyzed software can be mapped to the proposed service description through an intermediate mapping layer.

The proposed model of a service consists of the following:

- Description—a general description of the service with its name, identifier etc.

- Functional—contains descriptions of functions performed by the service.

  - Semantic—a semantic description of the service in terms of concepts expressed in a natural language.

    - IOPE—contains a description of inputs, generated outputs, preconditions and results associated with the service.
    - Category—contains a software group that describes the service.

  - Technical—details of how the service should be invoked.

    - Middleware—characterizes the middleware used to invoke the service; can be a software or human layer.
    - Launch control—contains details of how the service needs to be called.
      - Control—specifies whether the service is queued, when and with what parameters it is run.
      - Address—contains the location of the service including a system-wide identifier of the resource on which the service is run (such as a cluster or a server) and the service location within the resource.
    - Data management—describes how the service can handle data and constraints associated with the data.
    - Notification—defines when and how the user can be notified about events associated with the service.
      - Failure—defines how the user is notified about failure of the service.
      - Success—defines how the user is notified about successful completion of the service.
    - Technology details—a container for specific data required for the particular service type.

- QoS—describes quality metrics that can be defined and/or measured for the service.

- Reliability—indicates how reliable the results produced by the service are.
- Availability—presents how accessible the service is, which characterizes the network connecting to the service as well.
- Conformance—shows whether the service can really accept the data types it declares as its input and whether the output data match declared output data types.
- Green computing—defines parameters of the service important for energy management and corresponding startup of the service.

Models of resources and middleware are shown in Figure 3.3. As the generalized service description allows us to define services offered through various middleware, Figure 3.3 also presents the most common values for middleware and middleware code. Allowed values for the type of middleware are marked with * while allowed values for middleware code are marked with **.

Furthermore, meaning and types of values for all the leaves of the tree in the tree-like service description are explained in Tables 3.1 and 3.2.

Some of these quality metrics can change in time. For example:

- Cost [224] that can change depending on the time of day and may trigger even migration of computations from services on one hemisphere to another as shown in Section 7.2.

- Availability [226, 227, 172, 48], accessibility [172], reputation [226], and reliability [223].

This implies the need for learning about services [77].

3.2   Uniform Description of Modern Distributed Systems

Along the generalized service model, the author proposes a general description of a distributed system hosting that service. It is a uniform description of any hosting platform for any of the previously described service types and can also store specific technological details of the given platform. The platform is called a *resource* and has the following two main attributes:

- Description—for storing information about the resource and its reference.

- Access—contains both the address of the platform and registered *middlewares* through which services can be run on the platform. For each middleware, proper credentials need to be provided.

Table 3.3 lists parameters of the resource. Each middleware can be described by the technical aspects listed in Table 3.4.

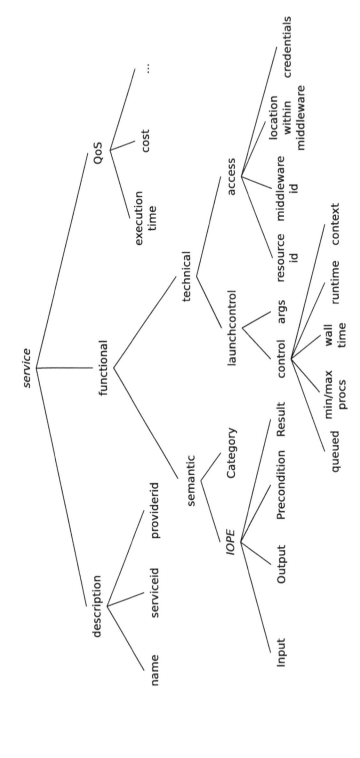

FIGURE 3.1: General service description 1/2.

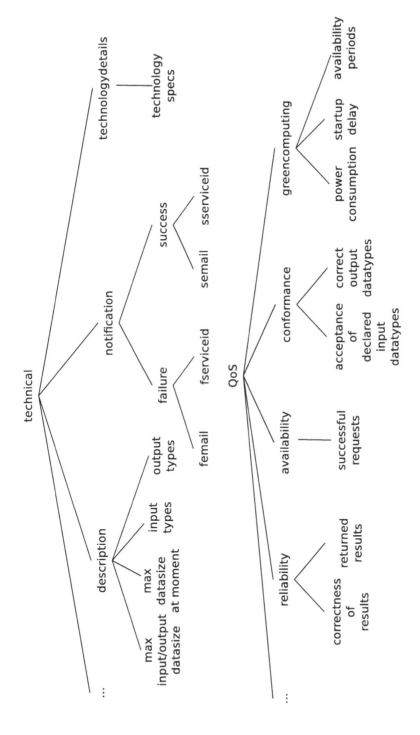

FIGURE 3.2: General service description 2/2.

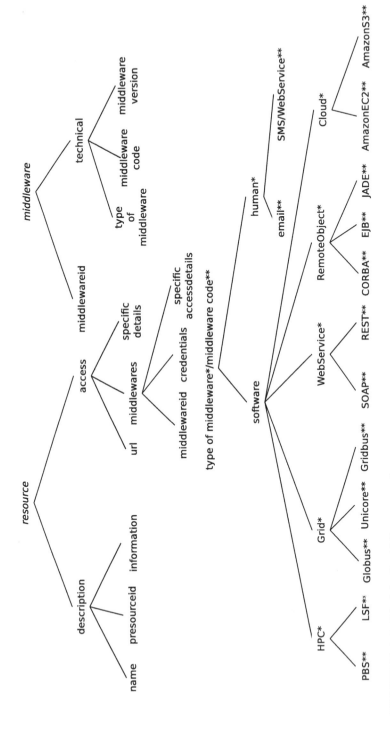

FIGURE 3.3: Model of middleware and resource.

TABLE 3.1: Model of a service 1/2.

QoS Metric	Data Type	Description
name	String	Service human-friendly ID
serviceid	long	ID of the service
providerid		ID of the service provider
Input	String[]	An array of concepts that describe inputs expressed in the natural language
Output	String[]	An array of concepts that describe outputs expressed in the natural language
Precondition	String[]	An array of concepts that describe conditions that must be met before the service is invoked expressed in the natural language
Result	String[]	An array of concepts that describe results associated with the service
Category	String[]	An array of concepts that describe categories to which the service can be assigned expressed in the natural language
middlewareid	long	Identifies the middleware that is used to access the service
min/maxprocs	long[]	Minimum and maximum numbers of processors on which the service can be invoked
walltime	long	The maximum expected running time of the service
runtime	long	The particular moment of time when the service should be run, if specified
context	String[]	Technical description of the conditions which must be met before the service is invoked, e.g., appearance of a file in a given location
presourceid	long	An ID of a physical resource such as a cluster or server on which the service has been installed
location	String	Can be a file path for an executable on an HPC cluster or a URL of a Web Service
args	String[]	Arguments given to the service (other than files passed within input files)
executiontime	double[]	Execution time declared by the provider over time
cost	double[]	Cost of the service declared by the provider over time
correctness of results	int[]	Whether results returned were correct over time

TABLE 3.2: Model of a service 2/2.

QoS Metric	Data Type	Description
returned results	int[]	Whether the service has returned results for input over time
successful requests	int[]	Whether the resource on which the service had been installed could be reached over time
acceptance of declared input datatypes	int[]	Whether the declared datatype has been accepted by the service over time
correct output datatypes	int[]	Indicates whether output data types correspond to the declared ones in field ** over time
power consumption	float	Declared by the provider per a particular size of input data
startup delay	float	Indicates how long it takes for the resource on which the service has been installed to resume operation and be ready to invoke the service
availability periods	Period long;long []	Indicates when the service is available without additional delay
max input/output datasize	double[]	Maximum allowed input and produced output data types
max data size at moment	double	Maximum data size accepted at each moment in time for for all instances started with various input data by the given client
inputtypes	String[]	Concrete file types accepted by the service
outputtypes	String[]	Concrete output file types produces by the service
femail	String	The email used to notify about failure of the service
fserviceid	SNotification {long;String[]} []	The ID and arguments of the service which should be invoked in case of service failure
semail	String	The email used to notify about success of the service
sserviceid	SNotification {long;String[]} []	The ID and arguments of the service which should be invoked after the service has been completed successfully
technology specs	byte[]	Specific parameters important for the given service type

3.3   Mapping of Various Service Types and Legacy Software to the Generalized Service

Table 3.5 gives a general mapping of the particular types of software listed in Section 2.1 to the proposed service, middleware, and resource descriptions.

TABLE 3.3: Model of a resource.

Resource Parameter	Data Type	Description
name	String	Name of the resource
presourceid	long	ID of the resource
information	String	Additional description of the resource
URL	String	The address where the resource can be accessed
specificdetails	String	Details related to the access to the resource (additional parameters, etc.)
middlewareid	long	Identifies the middleware that is used to access the service (there may be several middleware per resource)
credentials	String	Needed to pass to the middleware to access the resource through it
specificaccessdetails	String	Details needed to pass to the middleware to access the resource

TABLE 3.4: Model of middleware.

Middleware Parameter	Data Type	Description
middlewareid	long	Identifies the middleware that is used to access the service
type of middleware	String	Identifies the particular type of middleware (allowed values are listed with * in Figure 3.3)
middleware code	String	Identifies a particular software package of the middleware (allowed values are listed with ** in Figure 3.3)
middleware version	String	The particular version of the middleware specified in field middleware code

Mapping to the three previously identified components (service, middleware and resource) have been proposed. Additionally, details of particular mappings can be expressed by assignment of the former to particular fields of the components as outlined in Tables 3.1, 3.2, 3.3, and 3.4.

TABLE 3.5: Mapping of software to services.

Software	Service	Middleware	Resource
HPC	Executable	Queuing system (PBS, LSF etc.)	Cluster / server / workstation

Web Service	Service	Service type / server, e.g., SOAP/REST / server name and version	Server
Components in multitier systems, e.g., EJB	The remote method(s)	EJB remote interface	Java EE server
CORBA object	Methods of the object OMG Interface Definition Language (IDL) interface	ORB	The server on which the object is located
JADE agent	Programmed agent's behavior	JADE platform [42] / ACLMessages	Server on which the agent is running
Grid service	Service, e.g., a WSRF service (Globus 4.x)	e.g., Globus Toolkit	Server on which the grid service was deployed or automatically selected by the middleware
	Proxy Web Service calling GRAM	Globus through GRAM	Server on which the proxy service was deployed
Volunteer computing software	Web Services set up to define a project, upload input data, and browse results (Internet clients perform computations)	Comcute [37, 75]	Server with the access Web Services
	Proxy Web Service exposing operations to define a project, upload input data, and browse results, the service communicates with the system using the HTTP protocol [163]	BOINC [30]	Server with the proxy Web Service

Cloud service	Code of the service	Cloud interface, e.g., Amazon EC2	Cloud access server made available by the provider
Mobile software version 1	A proxy Web Service that accepts input data to a mobile device and allows browsing results; the mobile device accesses the proxy	Service type / server, e.g., SOAP/REST / server name and version	Server with the Web Service
Mobile software version 2	A service on the mobile device connected to the network	Service type, e.g., SOAP/REST	The mobile device

3.4 A Solution for Dynamic QoS-Aware Optimization of Service-Based Workflow Applications

3.4.1 Proposed Architecture

The following layers with corresponding modules are proposed by the author to address and solve the identified challenges in distributed QoS-aware service integration in distributed systems (Figure 3.4):

- *Workflow editor and management interface layer*—the layer allows you to compose complex tasks out of simple ones as well as specify functional and QoS requirements on the tasks such that services can be assigned to them. Alternatively, specific services can be assigned to the tasks by the user. Additionally, the layer allows you to manage both the complex tasks (workflow applications) as well as instances of the complex tasks run; it includes:
  - workflow editor,
  - workflow instance management system.

- *Service repository and workflow optimization layer*—the layer stores information about services registered on the platform (service repository) and is able to perform selection of services for a complex task based on the functional requirements expected from the services, the registered services, and QoS evaluation based on the service data (may change in time) and the evaluation of the environment parameters such as the network; it includes the following modules:

- service repository,
- dynamic workflow optimization.

- *Dynamic monitoring and runtime execution layer*—this layer is responsible for dynamic monitoring of service and their parameters and storage of this information in the service repository. Additionally, a dedicated module performs real-time management of workflow execution. The latter can be performed by either an engine (possibly clustered) that calls particular services or distributed software agents utilizing knowledge gained about the environment at runtime to migrate agents to such locations as to minimize communication costs. The latter can also control, e.g., the workflow data footprint. The following modules are included:

  - service monitoring,
  - dynamic distributed workflow execution.

- *Service adapter layer*—this layer contains plugins (adapters) that translate invocations to the generalized service to the API of the actual software system and transforms output from the system to the output of the generalized service. The following adapters are proposed:

  - Web Service to service adapter,
  - HPC queuing system to service adapter,
  - grid middleware to service adapter,
  - cloud API to service adapter,
  - mobile device proxy to service adapter, and
  - any other software from Section 2.1 to service adapter.

- *Distributed software system layer*—this is the layer of the actual distributed systems analyzed in Section 2.1. At this level, each of the systems has its own API and access methods.

Furthermore, Figure 3.5 presents assignment of services from various fields and technologically available through distinct middleware to particular workflow tasks. The concept allows integration of such services into complex scenarios spanning business and scientific worlds.

3.4.2   A Generalized Dynamic Workflow Scheduling Model

A workflow application is defined as a Directed Acyclic Graph (DAG) $G(T, E)$ in which vertices $t_i \in T$ correspond to simple tasks while the directed edges denote dependencies between the tasks as shown in Figure 3.6.

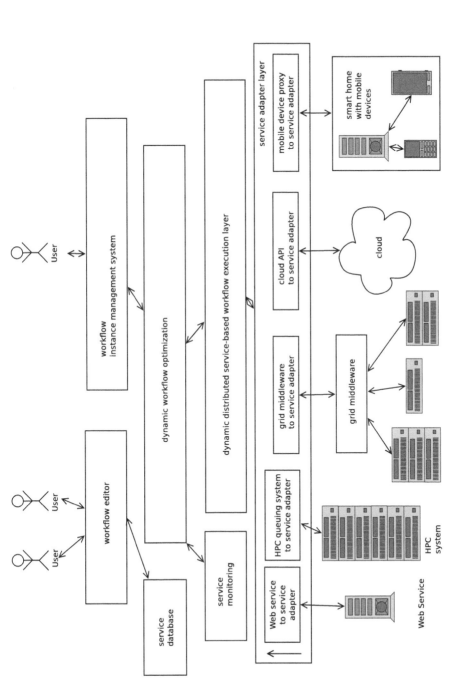

FIGURE 3.4: Proposed architecture for a dynamic distributed service-based multidomain workflow management system.

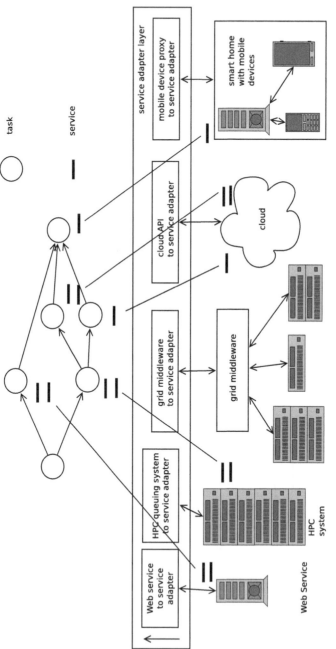

FIGURE 3.5: Workflow and service-to-task assignment.

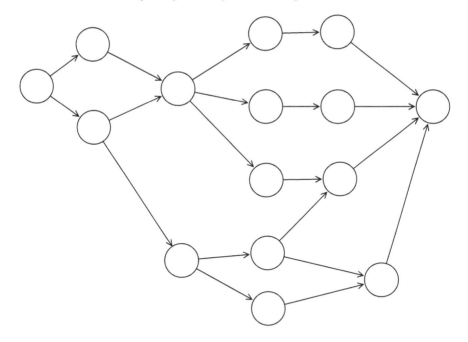

FIGURE 3.6: Generalized workflow model.

The proposed workflow model introduces several aspects identified in Section 2.2 at the workflow definition level. Namely, each workflow task is assigned the following attributes:

- Data handling mode:
  - Batch—data to a task may be supplied in a single data set.
  - Streaming—the task may accept ongoing data sets, possibly of various sizes.

- Data partitioning mode:
  - Fxed—the size of data processed by the task is set a priori.
  - Automatic data parallelization—an algorithm may decide how data is partitioned among workflow tasks, which may be useful for data parallelization and minimization of the workflow execution time.

In subsequent steps, the workflow scheduling problem formulations are presented.

3.4.3   A QoS-Aware Workflow Scheduling Problem

The problem can be defined as follows:

Input data:

1. the workflow graph $G(T, E)$ with batch data processing tasks,

2. sets of services $S_i = \{s_{i1}, s_{i2}, \ldots, s_{i|S_i|}\}$ for each task,

3. quality metrics q_{ij}^k of service s_{ij},

4. data sizes d_i to be processed in particular tasks t_i.

The goal is to find an assignment $\forall_i \, t_i \rightarrow (s_{i\,sel(i)}, t_i^{st})$ (where $sel(i)$ denotes the index of the service selected for task t_i and **sel** denotes a vector of these selections) such that execution of services on particular resources does not consume more resources than available (e.g., more processing cores than available) and a given QoS function is minimized:

$$min \, f_o(\mathbf{Q}, \mathbf{d}) \tag{3.1}$$

where:

- $\mathbf{q^k}$ is a vector of the k-th quality metric for the selected service $sel(i)$ for all workflow tasks T $\mathbf{q^k} = [q_{1\,sel(1)}^k, q_{2\,sel(2)}^k, \ldots, q_{|T|\,sel(|T|)}^k]^T$; matrix $\mathbf{Q}$ can then be defined as:

$$\mathbf{Q} = [\mathbf{q^1 q^2 \ldots q^K}] \tag{3.2}$$

  where K denotes the number of quality metrics considered.

- $\mathbf{d}$ is a vector that denotes what data sizes are processed in particular workflow tasks $\mathbf{d} = [d_1, d_2, \ldots, d_{|T|}]^T$

and an array of constraints is satisfied:

$$g_o(\mathbf{Q}, \mathbf{d}) \leq G_0 \tag{3.3}$$

$$\ldots$$

$$g_m(\mathbf{Q}, \mathbf{d}) \leq G_M. \tag{3.4}$$

An example of practical formulations include minimization of the workflow execution time with a constraint on the total cost of selected services, i.e.:

- $q_{ij}^0 = c_{ij}$ where c_{ij} is the cost of service s_{ij} used for processing a unit of input data.

- $q_{ij}^1 = t_{ij}$ where t_{ij} is the workflow execution time of service s_{ij} for the unit of data.

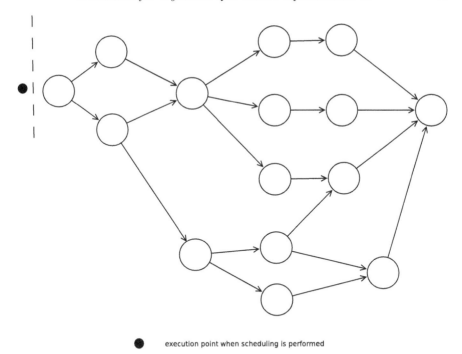

execution point when scheduling is performed

FIGURE 3.7: Scheduling with batch data handling tasks.

Then, the goal is to find such assignment of services running at particular points in time for each workflow task t_i such that the minimum workflow execution time is achieved:

$$min \ t_{\text{workflow}} \tag{3.5}$$

with a constraint that the total cost of used services must be below the available budget B:

$$\sum_i c_{i\ sel(i)} d_i \leq B. \tag{3.6}$$

Alternatively, other goals can be stated such as minimization of the total cost:

$$min \ \sum_i c_{i\ sel(i)} d_i \tag{3.7}$$

with a constraint on the workflow execution time:

$$t_{\text{workflow}} \leq T. \tag{3.8}$$

Assuming certain units of the quality metrics, other functions of time and cost may be minimized such as:

$$min \; t_{\text{workflow}} \cdot \left(\sum_i c_{i \; sel(i)} d_i \right) \tag{3.9}$$

depending on the required goal.

In fact, usually additional functions for aggregation of quality metrics for the whole workflow can be introduced. Then the goal to be optimized $f_o()$ can be expressed as a function of those functions. For instance, extending the quality metrics $\mathbf{q^0}$ and $\mathbf{q^1}$ assumed above with additional ones:

- $q^2_{i \; sel(i)} = r_{i \; sel(i)}$ denotes the reliability of service $s_{i \; sel(i)}$,

- $q^3_{i \; sel(i)} = powc_{i \; sel(i)}$ denotes the power consumption of a resource in which service $s_{i \; sel(i)}$ is installed,

one can define functions $qfinal^k(\mathbf{q^k}, \mathbf{d})$ that calculate the final quality value of metric k for the workflow:

- $qfinal^0(\mathbf{q^0}, \mathbf{d}) = \sum_i c_{i \; sel(i)} d_i$.

- $qfinal^1(\mathbf{q^1}, \mathbf{d}) = t_{\text{workflow}}$ where t_{workflow} can be computed using the workflow graph:

$$t_{\text{workflow}} = t_A + max_{i : \nexists k (t_i, t_k) \in E} \{ t_i^{\text{st}} + t_{i \; sel(i)}^{\text{exec}} \} \tag{3.10}$$

$$t_k^{\text{st}} \geq max_{i : (t_i, t_k) \in E} \{ t_i^{\text{st}} + t_{i \; sel(i)}^{\text{exec}} \} \tag{3.11}$$

where t_A denotes the total running time of the scheduling algorithm that delays the execution of the workflow (the initial scheduling or in case a previously chosen service has failed).

- $qfinal^2(\mathbf{q^2}, \mathbf{d}) = min_{path : T_i^2 = \{t_1^2, t_2^2, ..., t_x^2\} : (t_i^2, t_{i+1}^2) \in E} \prod_{i=1}^{x} r_{i \; sel(i)}$.

- $qfinal^3(\mathbf{q^3}, \mathbf{d}) = \sum_i powc_{i \; sel(i)} d_i$.

Then the final quality goal $f_o()$ can be described as a composition of functions $qfinal^k (\mathbf{q^k}, \mathbf{d})$s e.g.:

$$f_o(\mathbf{Q}, \mathbf{d}) = qfinal^1(\mathbf{q^1}, \mathbf{d}) \tag{3.12}$$

with additional constraints such as:

$$g_0(\mathbf{Q}, \mathbf{d}) = qfinal^0(\mathbf{q}^0, \mathbf{d}) \leq B$$
$$g_1(\mathbf{Q}, \mathbf{d}) = -qfinal^2(\mathbf{q}^2, \mathbf{d}) \leq -R$$
$$\text{corresponding to}$$
$$qfinal^2(\mathbf{q}^2, \mathbf{d}) \geq R$$
$$g_2(\mathbf{Q}, \mathbf{d}) = qfinal^3(\mathbf{q}^3, \mathbf{d}) \leq PD$$
$$g_0(\mathbf{Q}, \mathbf{d}) = qfinal^0(\mathbf{q}^0, \mathbf{d}) \leq B \qquad (3.13)$$

where $B > 0$, $R > 0$ and $PD > 0$.

Alternatively, the problem can be written as minimization of:

$$f_o(\mathbf{Q}, \mathbf{d}) = qfinal^0(\mathbf{q}^0, \mathbf{d}) \qquad (3.14)$$

with a constraint on the workflow execution time:

$$g_0(\mathbf{Q}, \mathbf{d}) = qfinal^1(\mathbf{q}^1, \mathbf{d}) \leq T \qquad (3.15)$$

or a product of execution time and cost:

$$f_o(\mathbf{Q}, \mathbf{d}) = qfinal^0(\mathbf{q}^0, \mathbf{d}) \cdot qfinal^1(\mathbf{q}^1, \mathbf{d}). \qquad (3.16)$$

This static problem in which scheduling occurs before the workflow starts (Figure 3.7) and for which all services are available for the duration of workflow execution is further extended by the author into a definition of a dynamic problem.

3.4.4 A Dynamic QoS-Aware Workflow Scheduling Problem

In this case, the following input data is assumed:

1. The workflow graph $G(T, E)$ with batch data processing tasks.

2. Sets of services $S_i(t) = \{s_{i1}, s_{i2}, \ldots, s_{i|S_i|}\}$ for each task; these sets can change over time.

3. Quality metrics $q_{ij}^k(t)$ of service s_{ij}; values of particular metrics can change over time as providers can change parameters as do providers on known markets.

4. Data sizes d_i to be processed in particular tasks t_i.

Change of sets S_i or quality metrics may appear due to (Figure 3.8):

1. Failure of a service at runtime and need for reselection of the next best service in terms of the defined QoS goals.

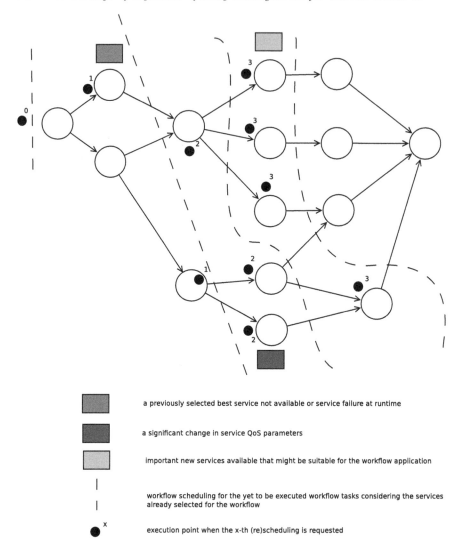

FIGURE 3.8: Scheduling and rescheduling of a workflow application with batch data handling tasks.

2. Temporary unavailability of the previously selected service for the task to be executed; this may be due to either network problems or a hosting issue of the platform on which the service is installed.

3. A significant change in service parameters that may make it worthwhile to reselect services even though this process takes time.

4. Showing up new potentially better services in terms of the defined global QoS goal.

This implies that rescheduling of services must be performed at various points of workflow execution as noted in Figure 3.8. Either one algorithm A can be rerun several times or various algorithms $A_1, A_2, \ldots, A_P$ will be run at the points where needed. For instance, a more complex algorithm can be run first and a coarser but faster algorithm can be launched later where only a few of the previously available services or their parameters have changed. Consequently, this brings the following definitions depending on time:

- $\mathbf{q^k}(t)$ is a vector of the k-th quality metric for the selected service $sel(i)$ for all workflow tasks T $\mathbf{q^k}(t) = [q^k_{1\ sel(1)}(t), q^k_{2\ sel(2)}(t), \ldots, q^k_{|T|\ sel(|T|)}(t)]^T$; matrix $\mathbf{Q(t)}$ can then be defined as:

$$\mathbf{Q(t)} = [\mathbf{q^1(t)q^2(t)} \ldots \mathbf{q^K(t)}] \qquad (3.17)$$

  where K denotes the number of quality metrics considered. Additionally, all points in time $t^*_1, t^*_2, \ldots, t^*_{MP}$ can be distinguished at which at least one value of $\mathbf{Q(t)}$ has changed. This can trigger a check on whether a scheduling algorithm should be rerun or not. In the latter case, expected benefits are lower than the cost by the presumably considerable execution time of the algorithm. Additionally, matrices $\mathbf{Q^k(t)}$ are introduced that hold the values of $q^k(t)$ for all values $t^*_i \le t$:

$$\mathbf{Q^k(t)} = [\mathbf{q^k(t^*_1)} \ldots \mathbf{q^k(t^*_x)}] : t^*_x \le t, \ t^*_{i+1} \ge t^*_i, \ t^*_{x+1} > t. \qquad (3.18)$$

  This matrix will contain the current and all past values for the given quality metric. When computing the final quality metric $qfinal^k$ past values may be considered with a certain weight.

Then values of exemplary quality metrics can be computed as follows at any moment t during workflow execution (this might be considered for both the initial scheduling and rescheduling):

- $qfinal^0(\mathbf{Q^0(t)}, \mathbf{d}) = \sum_{i:\text{not yet executed}} c_{i\ sel(i)}(t)d_i$
  $+ \sum_{i:\text{already executed}} c_{i\ sel(i)}(t^{st}_i)d_i$ where t^{st}_i denotes the time when task t_i was executed. After the service has been executed its quality metrics at that point in time are known now; in fact the value of the quality metric just before the service was executed can be set to reflect the measured value:

$$q^k_{i\ sel(i)}(t^{st}_i) = q^k_{i\ sel(i)}(t^*_x) : t^*_x \le t^{st}_i, \ t^*_{x+1} > t^{st}_i \qquad (3.19)$$

- $qfinal^1(\mathbf{Q^1(t)}, \mathbf{d}) = t_{\text{workflow}}$ where t_{workflow} can be computed recursively:

$$t_{\text{workflow}} \quad = \quad max_{i:\nexists k(t_i,t_k)\in E}\{t_i^{\text{st}} + t_{i\ sel(i)}^{\text{exec}}\} \qquad (3.20)$$

$$t_k^{\text{st}} \quad \geq \quad t_{A(t_i^{\text{st}})} + max_{i:(t_i,t_k)\in E}\{t_i^{\text{st}} + t_{i\ sel(i)}^{\text{exec}}\}. \qquad (3.21)$$

where $t_{A(t_i^{\text{st}})}$ denotes the time of the (re)scheduling algorithm run before task t_y which can be the measured or estimated; if not run it is assumed to be 0. This formulation also includes the algorithm run before the workflow starts.

- calculating the minimum of reliabilities for workflow path execution can be done considering measured (known) reliability at time t_i^{st}:

$$qfinal^2(\mathbf{Q^2(t)}, \mathbf{d}) = min_{path:T_i^2=\{t_1^2,t_2^2,...,t_x^2\}:(t_i^2,t_{i+1}^2)\in E}$$

$$\prod_{i=1}^{x}((t_i \text{ already executed})?r_{i\ sel(i)}(t_i^{\text{st}}) : r_{i\ sel(i)}(t)), \qquad (3.22)$$

- calculating the exemplary power-aware metric dependent on processed data sizes (useful when service execution time is a linear function of input data size) can be done considering the power consumption measured at time t_i^{st}:

$$qfinal^3(\mathbf{Q^3(t)}, \mathbf{d}) = \sum_i ((t_i \text{ already executed})?powc_{i\ sel(i)}(t_i^{\text{st}})d_i$$
$$: powc_{i\ sel(i)}(t)d_i).$$

3.4.5   A Dynamic QoS-Aware Workflow Scheduling Problem with Data Streaming

In this case, tasks marked with the streaming mode launch services that can process data packets as they come (Figure 3.9). As long as processing units such as processors or cores are available, this allows parallelization within the nodes on which services are installed, apart from potential data parallelization in parallel workflow tasks.

Let d_i^j denote the j-th data packet processed for task t_i. Obviously $\sum_j d_i^j = d_i$, which denotes the total size of data processed by task t_i. It should be noted that in case of rescheduling, various data packets processed in the same task can be processed by various services. This can be in case some packets have already been executed by a given service, the service has failed or become unavailable. In this case, an alternative service needs to be chosen. All in all, the following parameters can be distinguished in this model:

1. The workflow graph $G(T, E)$ with tasks processing data either in streaming or batch modes.

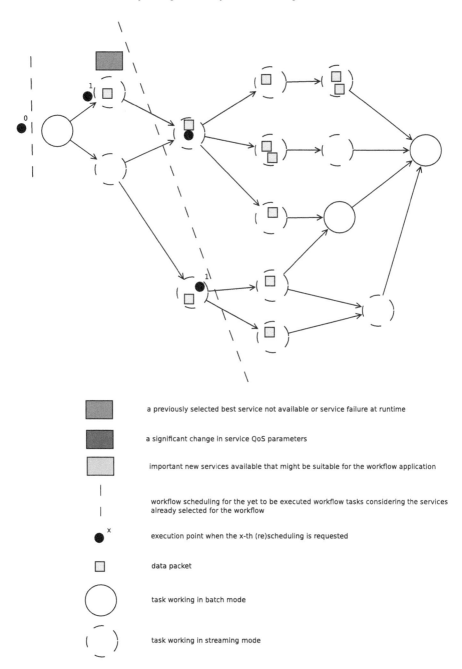

FIGURE 3.9: Scheduling and rescheduling of a workflow application with tasks handling data in streaming or batch modes.

2. Sets of services $S_i(t) = \{s_{i1}, s_{i2}, \ldots, s_{i|S_i|}\}$ for each task; these sets can change over time.

3. Quality metrics $q_{ij}^k(t)$ of service s_{ij}; values of particular metrics can change over time as providers can change parameters as do providers on known markets.

4. Data packets of sizes d_i^j to be processed in particular tasks t_i. It should be noted that the numbers of data packets for various tasks can be different as packets may be partitioned into smaller packets by services as the workflow is being executed. Let $\mathbf{d_i} = [d_i^1 d_i^2 \ldots d_i^{i\ \text{dcount}}]$ denote the vector of data packets for a particular task t_i.

Then, based on the considered sets of quality metrics, the following evaluation can be applied:

- Calculating the cost of workflow execution can be done considering data packets as follows:

$$qfinal^0(\mathbf{Q^0(t)}, \mathbf{d_1}, \ldots, \mathbf{d_{|T|}}) =$$

$$\sum_{i,j:\text{not yet executed data packet } d_i^j} c_{i\ sel(i)}(t)d_i^j +$$

$$\sum_{i,j:\text{already executed data packet } d_i^j} c_{i\ sel(i)}(t_i^{j\ \text{st}})d_i^j \qquad (3.23)$$

where $t_i^{j\ \text{st}}$ denotes the time when the j-th packet for task t_i was executed. After the service has been executed, its quality metrics at that point in time are known.

- $qfinal^1(\mathbf{Q^1(t)}, \mathbf{d_1}, \ldots, \mathbf{d_{|T|}}) = t_{\text{workflow}}$ where t_{workflow} can be computed as shown in Figure 3.10. Figure 3.10 considers tasks working in either streaming or non-streaming modes.

- calculating the minimum of reliabilities for workflow path execution can be done considering measured (known) reliability at time $t_i^{j\ \text{st}}$ for a data packet:

$$qfinal^2(\mathbf{Q^2(t)}, \mathbf{d_1}, \ldots, \mathbf{d_{|T|}}) = min_{path:T_i^2=\{t_1^2,t_2^2,\ldots,t_x^2\}:(t_i^2,t_{i+1}^2)\in E}$$

$$\prod_{i,j}((\text{packet } d_i^j \text{ already executed})?r_{i\ sel(i)}(t_i^{j\ \text{st}}) : r_{i\ sel(i)}(t)). \qquad (3.24)$$

- calculating the exemplary power-aware metric dependent on processed data sizes can be done considering the power consumption measured at time $t_i^{j\ \text{st}}$ for a data packet:

$$qfinal^3(\mathbf{Q^3(t)}, \mathbf{d_1}, \ldots, \mathbf{d_{|T|}}) =$$

$$\sum_{i,j}((\text{packet } d_i^j \text{ already executed})?powc_{i\ sel(i)}(t_i^{j\ \text{st}})d_i^j : powc_{i\ sel(i)}(t)d_i^j).$$

In order to compute t_{workflow} following and extending the approach proposed in [64], the following variables can be introduced to model calculation of the workflow execution time:

- $Succ(t_i)$—the set of tasks following task t_i in graph $G(T,E)$ i.e., $t_k \in Succ(t_i)$ if $\exists (t_i, t_k) \in E$,

- T_{start}—the set of workflow tasks that do not have predecessors,

- TH_i—the set of currently active threads related to task t_i,

- th_i^j—the j-th thread related to task t_i; it can be either related to copying a data chunk to service $s_{i\ sel(i)}$ or processing data.

Compared to [64], Figure 3.10 models both non-streaming and streaming tasks as well as rescheduling of the workflow if needed for particular tasks. The concept is similar to approaches with pushing data along edges of a directed graphs in maximum flow algorithms [203].

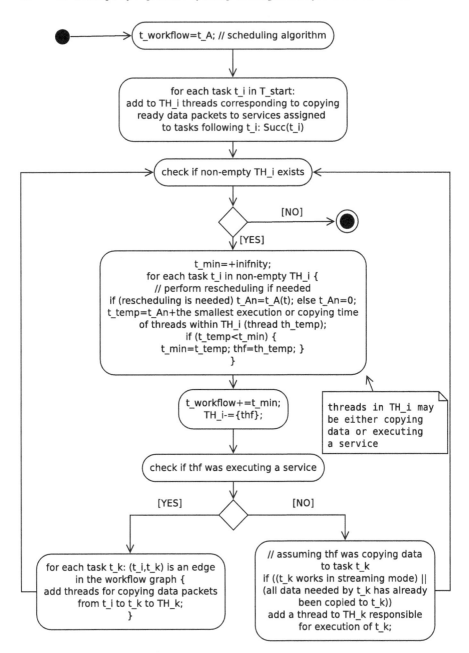

FIGURE 3.10: Evaluation of the workflow execution time, extending the concept from [64] with (re)scheduling as well as streaming and non-streaming modes.

Chapter 4

Dynamic Workflow Application Scheduling with Quality of Service

4.1   Monitoring and Measurement of Quality Metrics

4.1.1   Monitoring Procedure

The quality metrics identified and proposed in Section 3.1 need to be gathered for services, their providers, and the network that separates the client and the service. Moreover, in many cases the raw measured values should not be a basis for the values used in workflow application scheduling due to the following reasons:

1. Some values may be short-lived and may not reflect the actual values over a period of time.

2. Values do not reflect the rate of change of values either currently or in the past, which may be very important for some quality metrics such as cost or reliability.

3. Values of some quality metrics published by the provider, e.g., the cost may be subject to intentional manipulation in order to gain market share and further adjustment and possibly cause so-called vendor lock-in and losses to consumers.

In order to cope with these potential problems, the author proposed a quality measurement scheme depicted in Figure 4.1 extending the proposal from [65]. Technically, for each service s_{ij}, separate monitoring threads are invoked to check the quality metrics of corresponding services periodically with frequencies f^ks, possibly different for each quality metric k.

The frequencies f^k of measurements of various quality metrics may be either defined a priori or determined from execution of successive workflow applications.

- A priori—cost, availability

- When each workflow application involving the service is executed—execution time, reliability, cost, conformance, green computing

Each monitoring thread saves the data in a service repository from which the data can be fetched independently by an engine that performs service selection and scheduling for a particular workflow application. Writing and reading to and from the service repository is synchronized.

4.1.2   Measurement and Digital Filtering of Data

The values measured for particular quality metrics may be further filtered and processed before used by the scheduling engine. The following steps can be executed for each service s_{ij}:

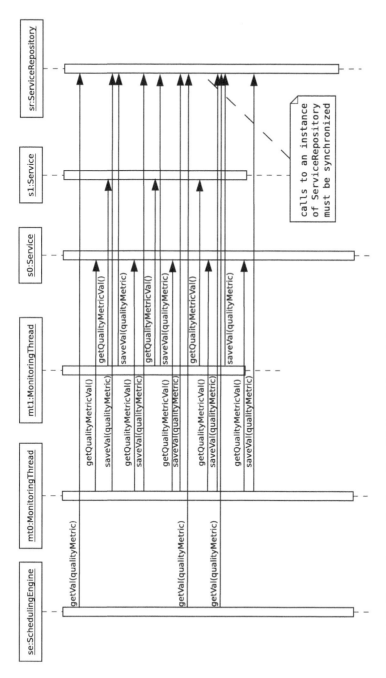

FIGURE 4.1: Dynamic monitoring of multiple quality metrics.

1. Registration of successive values $qr^k_{s_{ij}}(\frac{i}{f^k}) : i = 0, 1, \ldots$ in discrete time steps depending on the frequency of monitoring defined for the quality metric; for simplicity of notation we assume that $qr^k_{s_{ij}}[ind(t)]$ contains the value for time point closest to t where $ind(t) : R \to Z_+$ returns the index in an array for time t and $qr^k_{s_{ij}}[ind(t) - i]$ $i \in Z$ corresponds to the i-th value measured in the past from t with frequency f^k, i.e., at the point closest to moment $t - \frac{i}{f^k}$.

2. Storage of the measured values in a vector (can be stored as a running buffer) of a predefined length d^q: $\mathbf{qr^k_{s_{ij}}}(t) = [qr^k_{s_{ij}}[ind(t)]qr^k_{s_{ij}}[ind(t) - 1] \ldots qr^k_{s_{ij}}[ind(t) - (d^q - 1)]]^T$.

3. Computation of a function for the quality metric at a particular point in time based on the previously registered values from the running buffer, i.e.,

$$q^k_{s_{ij}}[ind(t)] = fq^k(qr^k_{s_{ij}}[ind(t)]).$$

In particular, fq^k might involve:

- changes of qr, in successive time points which can be depicted as $dqr^k_{s_{ij}}[ind(t)] = |qr^k_{s_{ij}}[ind(t)] - qr^k_{s_{ij}}[ind(t) - 1]|$,
- acceleration of qr, i.e., the speed of changes of qr which can be effectively modeled as $d2qr^k_{s_{ij}}[ind(t)] = |dqr^k_{s_{ij}}[ind(t)] - dqr^k_{s_{ij}}[ind(t) - 1]|$.

Table 4.1 presents exemplary functions fq^k that are suitable for certain groups of QoS metrics with corresponding descriptions. Not only can the past values be considered but also prediction of value changes in the future. Indeed, in case of workflow scheduling discussed throughout this chapter, scheduling occurs in fact before services are executed. The latter is true both in case of static scheduling and dynamic rescheduling when the remaining part of the workflow application must be scheduled again. In order to estimate the value of the quality metric at point $t^* > t$ when t denotes the current time, various approaches might be adopted, again depending on the type of the quality metric. For instance:

$$q^k[ind(t^*)] = qr^k[ind(t)] + (t^* - t)\frac{(qr^k[ind(t)] - qr^k[ind(t - a)])}{a}, \quad (4.1)$$

which predicts the value based on the past (a time before). For some quality metrics, such as the execution time, additional weight can be put on values that occurred at this particular time of day/week/month/year before, e.g., b weeks, c days and d hours ago. As an example, loads of workstations in computer laboratories may be similar to those at the particular time of the day on the given day of the week before.

It should be noted that depending on the quality metric, either small or large values are preferred. For example:

- small—cost, execution time, power consumption,
- large—reliability, availability, conformance.

TABLE 4.1: Exemplary functions fq^k.

Quality metric	Proposed fq^k	Description
Cost / Execution time / Power consumption	$q^k[ind(t)] = \frac{1}{x}\sum_{i=0}^{x} qr^k[ind(t)-i]$	a running average that is a low-pass filter that suppresses temporary highs and lows; it gives a better estimate of the value in the past but can omit sudden peaks
Cost / Execution time / Power consumption	$q^k[ind(t)] = \frac{1}{x}\sum_{i=0}^{x} qr^k[ind(t) - i] + \beta dqr^k[ind(t)]$	as above but considers sudden changes in the nearest past only as a negative, the far past is not considered
Cost / Execution time / Power consumption	$q^k[ind(t)] = \frac{1}{x}\sum_{i=0}^{x} qr^k[ind(t) - i] + \frac{\beta}{y}\sum_{i=0}^{y} dqr^k[ind(t) - i]$	as above but considers sudden changes in the defined window as a negative
Cost / Execution time / Power consumption	$q^k[ind(t)] = \frac{1}{x}\sum_{i=0}^{x} qr^k[ind(t) - i] + \frac{\beta}{y}\sum_{i=0}^{y} dqr^k[ind(t) - i] + \frac{\gamma}{z}\sum_{i=0}^{z} d2qr^k[ind(t) - i]$	as above but additionally considers acceleration in the value changes as negative
Cost / Execution time / Power consumption	$q^k[ind(t^*)] = \frac{1}{x}\sum_{i=0}^{x} qr^k[ind(t^*) - i] + \frac{\beta}{y}\sum_{i=0}^{y} dqr^k[ind(t^*) - i] + \frac{\gamma}{z}\sum_{i=0}^{z} d2qr^k[ind(t^*) - i]$	as above but considers prediction of values till the moment the service is planned to be executed
Reliability / Availability / Conformance	$q^k[ind(t)] = min_{i=0,\ldots,x} qr^k[ind(t) - i]$	the minimum (worst case) of the recent history (if x is small) and the far history is not considered, can consider a wide window with the past for large values of x
Reliability / Availability / Conformance	$q^k[ind(t)] = min_{i=0,\ldots,x}(qr^k[ind(t) - i] + \beta dqr^k[ind(t) - i])$	not only the value of the quality metric but its changes are considered as a minimum of a combination of both in the predefined window in the past

Furthermore, as suggested in [65] for at least some quality metrics such as the cost, one may need additional means to prevent the provider from using dumping techniques which results in the cost of a service being unusually low for a prolonged period of time. Such strategy might possibly be used to gain market share and result in the vendor lock-in for the clients of the service. In order to introduce other competitors in such a scenario, various techniques could be used to alter the value of $q^k[ind(t)]$:

1. Rotate the ordering of a few of the best offers [65] with a small probability.

2. The values offered by providers are modified within a small margin using confidence factors assigned to them.

Whichever version is chosen may also depend on the preferences of the client. For instance, the client may forgive the service provider to fluctuate prices in the far past and is only interested in the recent history. The rationale may be that the service may have changed the owner in the meantime. On the other hand, in case of service availability and reliability the whole service history may be of importance to some sensitive clients.

4.2   Workflow Scheduling Algorithms

Upon measurement and filtering of quality metrics, the latter can be used in workflow scheduling algorithms that determine assignment of services to workflow tasks.

This section discusses several algorithms that were either proposed in the literature or extended by the author for the purpose of dynamic scheduling of workflow applications in dynamic environments. In [68], the author detailed these algorithms and provided a comparison of performance benchmarked in BeesyCluster described in Chapter 6.1.

Firstly, the scheduling problem in either static or dynamic forms was presented in Sections 3.4.3 and 3.4.4. The formulation of this problem is a specific instance of the general problem modeled by the author in Section 3.4.3 for the purpose of discussion and comparison of algorithms. For the benefits of clarity but without losing generality, we consider one of the most typical problems of minimization of the workflow execution time with a bound on the total cost of selected services.

Let us assume the following variables and parameters:

- t_i—task i (represented by a node in the workflow graph) of the workflow application $1 \leq i \leq |T|$.

- $S_i = \{s_{i1} \ \dots \ s_{i|S_i|}\}$—a set of alternative services each of which can perform functions required by task t_i; only *one* service must be selected to execute task t_i $1 \leq i \leq |T|$.

- s_{ij}—the j-th service that is capable of executing task t_i.

- $c_{ij} \in R$—the cost of processing a unit of data by service s_{ij} $1 \leq i \leq |T|$ $1 \leq j \leq |S_i|$.

- P_{ij}—provider who has published service s_{ij} $1 \leq i \leq |T|$ $1 \leq j \leq |S_i|$.

- N_{ij}—node (cluster/server) on which the service s_{ij} was installed and runs $1 \leq i \leq |T|$ $1 \leq j \leq |S_i|$.

- $sp_N \in R$—speed/performance of node N.

- $d_i = d_i^{in} \in R$ —size of data processed by task t_i $1 \leq i \leq |T|$.

- $d_i^{out} \in R$—size of output data from task t_i $1 \leq i \leq |T|$.

- $f_{t_i}()$—a function that determines the size of output data based on the size of input data: $d_i^{out} = f_{t_i}(d_i^{in})$, $d_{ij}^{in} \in R$, $d_{ij}^{out} \in R$—sizes of input and output data processed and produced by service s_{ij} $1 \leq i \leq |T|$ $1 \leq j \leq |S_i|$,

- $d_{ijkl} \in R$—size of data passed from service s_{ij} to service s_{kl} $1 \leq i \leq |T|$ $1 \leq j \leq |S_i|$ $1 \leq k \leq |T|$ $1 \leq l \leq |S_k|$; data is passed through a network which entails communication costs.

- $t_{ij}^{exec}(d_{ij}^{in}) \in R$—execution time of service s_{ij}.

- t_{ijkl}^{comm}—communication time corresponding to passing of data of size d_{ijkl} from service s_{ij} to service s_{kl} $1 \leq i \leq |T|$ $1 \leq j \leq |S_i|$ $1 \leq k \leq |T|$ $1 \leq l \leq |S_k|$.

- $t_{N_{ij}N_{kl}}^{startup}$, $bandwidth_{N_{ij}N_{kl}}$—startup time and bandwidth corresponding to a communication link between nodes N_{ij} and N_{kl} $1 \leq i \leq |T|$ $1 \leq j \leq |S_i|$.

- $t_{ijkl}^{tr} \in R$—additional translation time required if output data from service s_{ij} is sent as input data to service s_{kl} if providers of the services are different $P_{kl} \neq P_{ij}$ $1 \leq i \leq |T|$ $1 \leq j \leq |S_i|$ $1 \leq k \leq |T|$ $1 \leq l \leq |S_k|$.

- $t_i^{st} \in R$—starting time for service $s_{i \ sel(i)}$ selected to execute t_i $1 \leq i \leq |T|$.

- $t_{workflow} \in R$—wall time for the workflow, i.e., the time when the last service finishes processing the last data.

- $B \in R$—budget available for execution of a workflow.

- q_{ij}^ks—possibly additional quality metrics for service s_{ij} such as reliability, power consumption of the node the service was installed on, availability, conformance, etc.; these would be followed with necessary constraints.

Scheduling of a service-based workflow application can be stated as follows:

- *Input data*: $G(T, E)$, t_i, S_i, c_{ij}, P_{ij}, N_{ij}, sp_n, $d_i = d_i^{\text{in}}$, f_{t_i}, d_i^{out}, $t_{N_{ij}N_{kl}}^{\text{startup}}$, $bandwidth_{N_{ij}N_{kl}}$, t_{ijkl}^{tr},

- *Variables* that describe a particular solution:

  - assignment of a particular $sel(i)$-th service to task t_i,
  - assignment of particular start time t_i^{st} to task t_i such that services running on the same resources do not consume more processing units (such as processors, cores) than available.

  This can be described as assignment $t_i \rightarrow (s_{i\ sel(i)}, t_i^{\text{st}})$ with the latter constraint met,

- *Goal*: minimization of the workflow execution time t_{workflow} with a bound on the total cost, i.e., $\sum d_{ij}^{\text{in}} c_{ij} \leq B$ or a different goal as outlined in Section 2.2.5, depending on needs.

It should be noted, though, that there can be more constraints imposed at this point, as modeled in general in Section 3.4.3. Such constraints can incorporate quality metrics particularly desirable for specific service types as described in Section 2.3.3.

4.2.1   Integer Linear Programming

Generally, the problem of *linear programming* (LP) aims at finding the vector $\mathbf{x}$ of real non-negative values such that the value of a certain linear objective function is maximized or minimized while a set of linear constraints expressed as inequalities must be satisfied. Formally, this can be written as:

$$\text{find the maximum of } c^T \mathbf{x} \tag{4.2}$$

while the following constraints are satisfied:

$$A\mathbf{x} \leq \mathbf{b} \tag{4.3}$$

$$\mathbf{x} \geq 0. \tag{4.4}$$

Vectors $\mathbf{b}$ and $\mathbf{c}$ as well as matrix $\mathbf{A}$ are coefficients given as input while the value of vector $\mathbf{x}$ is to be found to satisfy the above.

Linear Programming (LP) has wide applications from scientific to business applications including transport, telecommunication, etc. Several algorithms [102] have been proposed to solve LP including:

1. the simplex algorithm [157] proposed by Dantzig,

2. a polynomial-time algorithm proposed by Karmarkar [125],

3. Khachiyan's algorithm [32].

Integer Programming allows the values of vector **x** to be integers only. This makes the problem much more difficult to solve, in fact NP-hard because of the bounds in the constraints. *Mixed Integer Linear Programming (MILP/ILP)* allows for some of the vector **x** elements to be real numbers and restricts some to be integers and is also NP-hard. Approaches such as the following are used:

1. branch and bound that can return a feasible solution quite early in the search,

2. cutting plane algorithms that use linear programming relaxation and going toward integer solutions,

3. heuristic algorithms including genetic algorithms, simulated annealing, tabu search, ant colony-like optimization, etc.

Firstly, in order to represent the problem as a mixed integer linear programming problem, the variables need to be encoded as the **x** vector. Traditionally, in the service selection problem [27, 28, 226], variables x_{ij} are introduced with the following meaning:

$$x_{ij} = \begin{cases} 1 & sel(i) = j \\ 0 & sel(i) \neq j. \end{cases}$$

In order to enforce these constraints in the ILP formulation, the following constraints can be stated:

$$\forall_i \sum_j x_{ij} = 1 \tag{4.5}$$

$$\forall_{i,j} \ x_{ij} \in Z, \ x_{ij} \geq 0. \tag{4.6}$$

In the scheduling problem, also assignments of services to resources at particular points in time must be taken care of along with possible execution of various services on a certain resource.

4.2.1.1 Assignment of Service per Task and ILP Constraints on Data

Firstly, let us focus on data processing constraints. In this problem formulation, we assume that the sizes of data processed by particular workflow tasks are known in advance. Indeed, in reality this is often the case. For instance, in the following scenarios:

- *Business*—a production company knows how much resources it needs to order from its subcontractors to achieve the assumed production size and its own business goal; ordering various types of items from third parties may be modeled as separate workflow tasks; obviously planning production size may be subject of economic optimization, which is out of scope of this book.

- *Scientific*—a workflow processing multimedia content requires particular phases that possibly convert, normalize, and analyze frames of known size coming at a known rate.

Note, however, that especially in scientific computing one may wish to parallelize processing on input data (*data parallelism*) by splitting the latter and parallel processing by, e.g., various HPC systems. This can be modeled as separate parallel tasks in the workflow with either predefined sizes assigned to various systems. However, it is also possible to model the workflow in such a way that scheduling will decide which of the parallel tasks processes how much of the input data based, e.g., on the performance and consequently execution time on the given cluster. A solution to this problem is presented by the author in Section 4.4.

Generally, input data of size d_i^{in} is processed by service $s_{i\ sel(i)}$ and output data of size d_i^{out} is produced ($d_i^{\text{out}} = f_{t_i}(d_i^{\text{in}})$). Depending on the application, output data can be either smaller, of equal, or larger size than the input data:

- Image processing, sorting, matrix operations—often $d_i^{\text{out}} = d_i^{\text{in}}$.

- Comparison of data—output data can be a flag indicating the result.

- Modeling—based on parameters of a model, a complex model of a domain can be created for which $d_i^{\text{out}} >> d_i^{\text{in}}$.

Function f_{t_i} characterizes the task, not the service, since all services executing the task must take the same input and return same output data. As outlined above, the size of the output data from a service can be related to its input data depending on what the service provides, e.g., conversion of an image from the 16-bit to 8-bit depth, merging sorted arrays.

The problem solution will assign the known d_i to exactly one $d_{i\ sel(i)}^{\text{in}}$ and make all other values equal to 0, which corresponds to selection of just one service per task:

$$d_{ij}^{\text{in}} = \begin{cases} d_i & \text{sel(i)=j} \\ 0 & \text{sel(i)}\neq\text{j}. \end{cases}$$

In terms of ILP constraints, this can be achieved by stating that the sum of data processed by particular services for each task t_i does not exceed d_i

$$d_i = d_i^{\text{in}} = \sum_j d_{ij}^{\text{in}} \quad d_{ij}^{\text{in}} \geq 0 \tag{4.7}$$

and forcing each d_{ij}^{in} to be either 0 or d_i. This can be done by introducing an additional variable $b_{ij} \in Z$ for each service s_{ij} that defines whether service s_{ij} has been activated or not:

$$b_{ij} = \begin{cases} 0 & \text{if } d_{ij}^{\text{in}} = 0 \\ 1 & \text{if } d_{ij}^{\text{in}} = d_i. \end{cases} \tag{4.8}$$

Since only one service s_{ij} per task t_i can be selected, the following needs to hold true:

$$\forall_i \sum_j b_{ij} = 1. \tag{4.9}$$

Then, large enough constants U_{ij} can be introduced such that:

$$d_{ij}^{\text{in}} \leq b_{ij} U_{ij}. \tag{4.10}$$

If $b_{ij} = 0$, then corresponding $d_{ij}^{\text{in}} = 0$. For the given i and exactly one k $b_{ik} = 1$ which, along with equation 4.7, will force $d_{ik}^{\text{in}} = d_i$.

The same strategy can be applied for output data with the same b_{ij}s variables.

$$d_i^{\text{out}} = \sum_j d_{ij}^{\text{out}} \tag{4.11}$$

Consequently, the execution time of service s_{ij} can be modeled as:

$$t_{ij}^{\text{exec}} = f_{ij}^{\text{exec}}(d_i^{\text{in}}).$$

For instance, for simple operations on images, $t_{ij}^{\text{exec}} = \alpha_{ij} d_{ij}^{\text{in}}$. Since d_{ij}^{in} is a variable of the ILP formulation, then it may seem that only linear complexities for modeling of a service are possible. However, since d_is are assumed to be constants, this case can also be expressed in the ILP form even if $f_{ij}^{\text{exec}}()$ is not linear. In fact, t_{ij}^{exec} will be either 0 or a particular value known in advance depending on whether d_{ij}^{in} is 0 or d_i^{in}. Similar to the aforementioned constraints, b_{ij}s can be used to determine whether t_{ij}^{exec} has the value of 0 or $f_{ij}^{\text{exec}}(d_i^{\text{in}})$. This comes at the cost of additional constraints in the model.

4.2.1.2   Communication

Furthermore, depending on the workflow application, output data from a task may be distributed among input data of following tasks in various ways, i.e., how d_l^{in} depends on d_k^{out} if $(t_k, t_l) \in E$. For instance, a copy of output data of size d_i^{out} can be sent to each of the tasks following t_i or it may be split into parts for parallel processing by following nodes or partitioned in a different way. Depending on the services chosen for execution of two following tasks, e.g., $t_k, t_l : (t_k, t_l) \in E$, sizes of data transferred between services need to be obtained in order to determine actual communication costs that stem from the locations of the chosen services. For instance, the following constraints describe distribution of output data among following tasks:

$$d_i^{\text{out}} = \sum_{j,l;k:(v_i,v_k)\in E} d_{ijkl} \tag{4.12}$$

$$d_k^{\text{in}} = \sum_{j,l;i:(v_i,v_k)\in E} d_{ijkl}. \tag{4.13}$$

It is also possible to define a constraint that, e.g., each of the following tasks will receive data of same size from its predecessors:

$$\forall_{k:(v_i,v_k)\in E} D = \sum_{j,l} d_{ijkl}. \tag{4.14}$$

Alternatively, for a given t_i the following constraint will send a copy of output data from node t_i to each of the following nodes

$$\forall_{k:(v_i,v_k)\in E} \ d_i^{\text{out}} = \sum_{j,l} d_{ijkl}. \tag{4.15}$$

Such configuration depends highly on the application. Then communication costs can be described considering d_{ijkl} values. Theoretically, the communication cost in case of following services installed on various nodes is modeled using startup time, bandwidth, and the size of the data sent in the following way:

$$t_{ijkl}^{\text{comm}} = \begin{cases} 0 & \text{if } N_{ij} = N_{kl} \\ t_{ijkl}^{\text{startup}} b_{ijkl}^{\text{startup}} + \frac{d_{ijkl}}{bandwidth_{ijkl}} & \text{otherwise.} \end{cases}$$

$b_{ijkl}^{\text{startup}}$ is an integer variable that indicates whether the pair of services s_{ij} and s_{kl} is sending data between each other:

$$b_{ijkl}^{\text{startup}} = \begin{cases} 0 & \text{if } d_{ijkl} = 0 \\ 1 & \text{if } d_{ijkl} > 0 \end{cases}$$

and can be stated as in the ILP formulation:

$$b_{ijkl}^{\text{startup}} \geq \frac{1}{U_{ijkl}} d_{ijkl}$$
$$b_{ijkl}^{\text{startup}} \leq 1. \tag{4.16}$$
$$b_{ijkl}^{\text{startup}} \in Z$$

where U_{ijkl} is a constant that is large enough so that $\frac{1}{U_{ijkl}} d_{ijkl} < 1$ if $d_{ijkl} > 0$. The initial data size for the workflow can be used as the upper bound on d_{ijkl}. Indeed, if $d_{ijkl} = 0$, then $b_{ijkl}^{\text{startup}} = 0$ and if $d_{ijkl} > 0$, then $b_{ijkl}^{\text{startup}} = 1$. Obviously, t_{ijkl}^{comm} will be an element of the workflow execution time to be minimized subject to other constraints such as the cost of selected services.

Consequently the startup times and bandwidths as well as data sizes will be considered. On the other hand, the model increases the number of constraints in order to model communication precisely. It may be profitable to give up consideration of, e.g., startup times if communication of large data sizes is considered, e.g., in multimedia workflow applications processing very large data sizes.

4.2.1.3 Workflow Execution Time and Service Scheduling on Resources

Now the total execution time of the workflow application can be modeled in stages, each of which represents time dependencies of the workflow graph edges:

$$\forall_{i,k:(v_i,v_k)\in E}$$

$$t_k^{\text{st}} \geq t_i^{\text{st}} + \sum_j t_{ij}^{\text{exec}} + \sum_{j,l} t_{ijkl}^{\text{comm}} + \sum_{j,l} t_{ijkl}^{\text{tr.}} \tag{4.17}$$

Obviously, only one service is selected for execution of task t_i and only this service will send output data to following tasks, in this case t_k. Finally, workflow execution time t_{workflow} needs to be larger than the termination time of computations of any workflow task:

$$\forall_{i:\nexists_q(v_i,v_q)\in E} \; t_{\text{workflow}} \geq t_i^{\text{st}} + \sum_j t_{ij}^{\text{exec}} + \sum_{j,l} t_{ijkl}^{\text{comm.}} \tag{4.18}$$

It should be noted that some services may be installed on the same node. In that case, proper constraints must be introduced to ensure that no more than currently available computing units (processors, cores) are used at any time. There are several approaches that can be adopted at this point such as:

1. Solving the ILP formulation of the workflow scheduling problem, i.e., minimization of t_{workflow} with the previously mentioned constraints and identification of potential overlapping of services running on the same node and spreading the schedule to remove the overlapping. This may not necessarily yield the optimal solution, however, the solution to the original ILP problem will need to be stopped after a timeout due to the complexity for larger workflow applications.

2. Identification of potentially overlapping pairs of services before the ILP problem is solved. This requires setting

$$t_i^{\text{st}} = \sum_j t_{ij}^{\text{st}}$$

$$\text{where: } t_{ij}^{\text{st}} = \begin{cases} 0 & \text{if } d_{ij} = 0 \\ t_i^{\text{st}} & \text{if } d_{ij} > 0 \end{cases}$$

and adopting one of the following two alternative solutions:

(a) imposing sequences of services beforehand, e.g.,

$$t_{ij}^{st} + t_{ij}^{exec} \leq t_{kl}^{st,} \tag{4.19}$$

(b) imposing conditions for the ILP solver by setting proper spreading equations as proposed in [68], i.e.:

$$\left| t_{ij}^{st} + \frac{t_{ij}^{exec}}{2} - t_{kl}^{st} + \frac{t_{kl}^{exec}}{2} \right| \geq \frac{t_{ij}^{exec}}{2} + \frac{t_{kl}^{exec}}{2} \tag{4.20}$$

The latter one can be expressed as ILP constraints following the solution proposed in [9]. In fact, it introduces an additional integer variable which makes the problem more difficult. It might be preferable to set only several sequences instead of all possible and consequently reducing the complexity of the problem.

Finally, the problem can be defined as:

$$\text{minimize } t_{\text{workflow}} \tag{4.21}$$

subject to constraints 4.7–4.18 with a bound on the total cost, i.e., $\sum d_{ij}^{in} c_{ij} \leq B$ and 4.20 if services using shared machines are considered.

4.2.2   Genetic Algorithms

Listing 4.2 presents a general concept of a genetic algorithm that can be used to optimize schedules for a workflow application. It follows the typical flow of the genetic algorithm with optimization trials in successive time steps with copying of a few of the best chromosomes to the next iteration, crossover of chromosomes selected with probabilities proportional to their fitness functions, and mutation.

Implementation of the representation proposed by the author in [68] is shown in Listing 4.1. The first $|T|$ elements of array `solution` denote indexes of services selected for tasks, i.e., $0 \leq e_i < |S_i|$ $0 \leq i < |T|$. Initial values are generated using random selection of services for tasks, i.e., $e_i = random()\ mod\ S_i$. The next $|T|$ elements denote ordering of the $|T|$ tasks. If t_k, t_l are executed by services which were installed on a single processor, t_l is executed after t_k if and only if $e_i = k\ e_j = l\ :\ i < j\ i \geq |T|$. Implementation of a genetic solution to this problem is proposed in Listing 4.2. The `score()` function returns a value $\frac{\alpha}{t_{\text{workflow}}}$ if the cost constraint has been met. This way higher scores correspond to shorter execution times. Illegal solutions obtain scores of 0. Initial values are chosen randomly out of following tasks starting from the initial task. In the next step, its successors are added to the pool of the tasks from which in the next step one will be chosen randomly and the procedure is repeated.

Crossover, i.e., creation of new chromosomes, is based on a pair of already existing chromosomes, i.e., selection of a set of services for a part of tasks from one chromosome and another set from the other chromosome for the other part of tasks. Order of tasks is either generated or taken from one of the chromosomes with a certain probability.

Listing 4.1: Chromosome in the genetic algorithm.

```
1 typedef struct {
    int *solution; // an array −
    // the first |T| elements denote service indexes for tasks
        i.e., solution[i]=service index for tᵢ
    // next |T| elements denote task ids − and the order of
        execution of particular tasks
    double score;
6 } t_chromosome;
```

Listing 4.2: Genetic algorithm.

```
t_chromosome chromosomes[population_size],new_chromosomes[
    population_size];
generate_initial_population();
for(i=0;i<MAXITERATIONS;i++) {
4   for(j=0;j<population_size;j++) {
        chromosomes[j].score=score(chromosomes[j].solution);
    }
    sort(chromosomes); // sort the array of chromosomes by
        score
    // from best to worst
9   for(j=0;j<best_count;j++) { // copy a few
        // of the best ones
        new_chromosomes[j]=chromosomes[j];
    }
    // apply crossover
14  for(j=0;j<(population_size−best_count);j++) {
        // select a chromosome with selection probability
        // proportional to scores
        temp_chromosome_1=random(chromosomes,population_size);
        temp_chromosome_2=random(chromosomes,population_size);
19      pivot=random()%|T|;
        for(k=0;k<pivot;k++)
            new_chromosomes[best_count+j].solution[k]=
                temp_chromosome_1.solution[k];
        for(k=pivot;k<|T|;k++)
            new_chromosomes[best_count+j].solution[k]=
                temp_chromosome_2.solution[k];
24      probability=(float)random()/RANDMAX;
        if(probability<CROSS_THRESHOLD)
            for(k=|T|;k<2*|T|;k++)
                new_chromosomes[best_count+j].solution[k]=
                    temp_chromosome_1.solution[k];
        else if(probability<2*CROSS_THRESHOLD)
```

```
29              new_chromosomes[best_count+j].solution[k]=
                   temp_chromosome_2.solution[k];
           else generate_new_order(new_chromosomes[best_count+j]);
        }
        for(j=0;j<population_size;j++) {// apply mutation with a
           small probability
           if ((float)random()/RAND_MAX<MUT_THRESHOLD) {
34            t_index=random()%|T|;
              new_chromosomes[j].solution[t_index]=random()%|St_index|;
           }
           if ((float)random()/RAND_MAX<MUT_THRESHOLD2)
              generate_new_order(new_chromosomes[j]);
39      }
     }
```

Mutation for service selection is applied as follows. With probability MUT_THRESHOLD a random service for a randomly chosen task is selected. With probability MUT_THRESHOLD2 a new order of tasks is generated. For the following experiments, the following thresholds were used: CROSS_THRESHOLD=0.35, MUT_THRESHOLD=0.1, MUT_THRESHOLD2=0.02, similarly to the experiments performed in [68].

4.2.3   Divide and Conquer

The divide-and-conquer approach shown in Listing 4.3 uses a technique to partition the initial problem into a number of subproblems which are then solved separately and the results are merged into a final solution for the initial problem. Since the algorithms for the subproblems will only use local knowledge, the final solution will likely be suboptimal. It allows us to solve the subproblems using any method, even different methods for various subproblems if such methods are better for specific types of subproblems.

In this case, let us assume that the initial workflow application with $|T|$ number of tasks is partitioned into a certain number of graphs $G_i(T_i, E_i)$

$$\forall_{i,j:i \neq j} \; T_i \cap T_j = \emptyset, \quad T_1 \cup T_2 \cup \dots T_X = T$$

each of which is solved independently. Listing 4.3 presents details of the algorithm for workflow application scheduling using the divide and conquer approach.

Note that in the proposed solution, the algorithm divides the remaining budget and partitions it equally among the remaining problems (graphs G_i) to be solved. Only if the problem cannot be solved (presumably because the budget bound was too tight), then the budget is increased by a predefined margin. This scheme follows until a solution for the current subproblem can be found. Then the algorithm follows with the remaining problems. It is possible, though, that finding a solution to a subproblem will fail regardless of the budget set considering the remaining budget. This can happen especially toward the end of the main loop. One of the following remedies could be applied then:

1. Repeat the algorithm with a smaller margin; in this case the first sub-problems may finish within tighter budgets and later ones may have larger budgets available.

2. Decrease the initial `tempbudget` threshold; some subproblems may finish within budgets smaller than the one scaled proportionally given the number of tasks.

3. Terminate with a failure information, as the problem probably cannot be solved with the input data and the requested budget bound.

Listing 4.3: Divide-and-conquer algorithm for workflow scheduling.

```
divide_and_conquer_workflow_scheduling (G,B) {
    TP=the maximum number of tasks in a subproblem;
    problemcount=⌈ |T|/TP ⌉;
    partition graph G(T,E) into problemcount graphs Gᵢ(Tᵢ,Eᵢ)
    such that T₁ ∪ T₂ ∪ ... ∪ Tproblemcount = T
    and ∀ᵢ≠ⱼTᵢ ∩ Tⱼ = ∅;
    remainingbudget=B;
    remainingproblems=problemcount;
    for (; remainingproblems >0; remainingproblems−−) {

        tempbudget= remainingbudget/remainingproblems ;

        while (1) {
            result=solve problem with Gᵢ(Tᵢ,Eᵢ) to minimize the
                workflow execution time for Gᵢ with cost bound
                tempbudget;
            if (result==NULL) // failed e.g., because
            // of too tight cost bound
                if (tempbudget>=remainingbudget) // fail
                    return FAIL;
                tempbudget+=margin;
            else
                break;
        }
        remainingbudget−=tempbudget;
        if (remainingbudget <0)
            return FAIL;
    }
}
```

It should be noted that the first two solutions will increase the number of calls to solving subproblems and consequently the execution time of the divide-and-conquer algorithm.

Another interesting aspect is that the algorithms launched to solve the subproblem can vary depending on the subproblem type, especially the graph characteristics. As an example, an optimal algorithm may be launched for really small subproblems while heuristic ones are started for larger subproblems. The size of a graph can be measured, e.g., by the number of tasks and edges.

4.2.4   Hill Climbing

The general idea of this algorithm is first to find a legal solution to a problem and then change a single part of the current solution in successive iterations if it improves the goal. Finally, this procedure ends in a result that most likely is a local minimum or maximum. In this particular problem, one can first obtain a solution that meets the constraints and then try to substitute one of the previously selected services with a new one such that the constraints are still met and the global optimization criterion has been improved. Such an approach was presented in [186] and named GAIN. In the context of this work two implementations of this idea are presented in Listings 4.4 and 4.5.

The problem with the approach shown in Listing 4.4 in this context is the long execution time of the workflow evaluation in line 6 of Listing 4.4. While computing the sum of the services selected for the workflow with the new service is fast, computing the execution time requires parsing the whole workflow graph. It is possible to simplify assessment of the quality with the new service at the cost of some accuracy which is considered as GAIN1 concept presented in [186]. In this version, evaluation of the whole workflow has been replaced with relative comparison of services. Pseudocode for this version is presented in Listing 4.5. Obviously, the local evaluation of `tempGainWeight` is now much faster than in the previous example. Tests of these two approaches for the workflow application scheduling indeed suggest that the latter one offers better performance overall despite theoretically more coarse evaluation but also with much shorter execution time of the algorithm.

4.3   Characteristics and Comparison of Algorithms

Naturally, the question arises as to which of the proposed algorithms should be preferred for a particular workflow application. Figure 4.2 presents characteristics of the algorithms in terms of:

1. Type of knowledge and view on the workflow application, which can be either local, e.g., in the case of divide-and-conquer where views of partitions are used to solve them individually and combine into final solution. Similarly, the optimized GAIN version also takes into consideration local metrics of individual services rather than for the whole workflow like

Listing 4.4: Algorithm based on the GAIN2 [186] concept.

```
prevService[t₁,t₂,...,t|T|]=the set of the cheapest services,
    one for each task;
while (1) {
    bestGainWeight=−∞;
    for (each workflow task tᵢ)
        for (each service sᵢⱼ) {
            newWorkflowEval=EvaluateWorkflow(prevService,tᵢ,sᵢⱼ);
            if (newWorkflowEval.getRunCost()>budget) continue;
            tempGainWeight=(double)(prevWorkflowEval.getRunTime()
            −newWorkflowEval.getRunTime())/(double)(1000+
            newWorkflowEval.getRunCost()−prevWorkflowEval.
                getRunCost());

            if (tempGainWeight>bestGainWeight) {
                bestService=sᵢⱼ;
                bestTask=tᵢ;
                bestGainWeight=tempGainWeight;
            }
    if (bestService!=prevService[bestTask])
        prevService[bestTask]=bestService;
    else break;
}
```

Listing 4.5: Algorithm based on the GAIN1 [186] concept.

```
prevService[t₁,t₂,...,t|T|]=the set of the cheapest services, one
        for each task;
while (1) {
    bestGainWeight=−∞;

    for (each workflow task tᵢ)
        for (each service sᵢⱼ) {
            if (total cost of selected services with sᵢⱼ>budget)
                continue;
            tempGainWeight=(double)(prevService[tᵢ].getRunTime()
            −sᵢⱼ.getRunTime())/(double)(1000
            +sᵢⱼ.getRunCost()−prevService[tᵢ].getRunCost());
            if (tempGainWeight>bestGainWeight) {
                bestService=sᵢⱼ;
                bestTask=tᵢ;
                bestGainWeight=tempGainWeight;
            }
    if (bestService!=prevService[bestTask])
        prevService[bestTask]=bestService;
    else break;
}
```

in the original version. On the other hand, the GA and ILP approaches clearly use knowledge about the whole workflow.

2. Type of the solution, i.e., heuristic by design or potentially optimal. Algorithms such as divide-and-conquer as well as ILP can return optimal solutions for small problems if a large enough timeout is set. Solving it optimally for large problems is unrealistic. Consequently, ILPHEU denotes ILP when run using a solver with a timeout set for a solution. GAIN and GA were designed as heuristic algorithms able to provide reasonable solutions in short time frames.

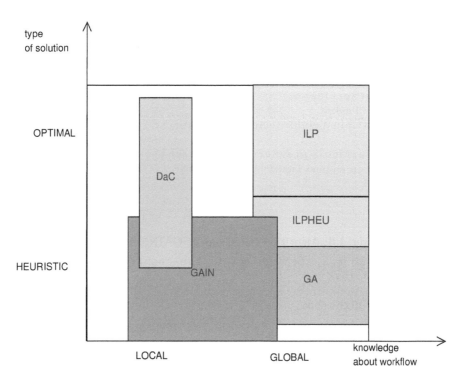

FIGURE 4.2: Characteristics of the algorithms.

In [68] the author compared the outlined algorithms for both balanced and irregular workflow applications with various sizes from 6 tasks and 30 services per workflow up to over 100 tasks and 400 services. Furthermore, [68] presents results for various availability of services. Before the workflow is started, static scheduling is performed including selection of services for particular tasks and assignment at particular time points. In case services fail with a certain probability, the total execution time of the workflow application increases because the algorithm needs to be rerun to select alternative services and considering the services already executed for the workflow.

Let us introduce the following notation for workflow graphs:

1. A→B—execution of part A of a workflow followed by execution of part B.

2. A||B—parallel execution of parts A and B.

3. A^n—parallel execution of n A parts.

4. nA—sequential execution of n A parts.

5. T1→T2—additional dependence of task T2 on task T1.

In order to assess relative performance of the algorithms, several tests have been performed in the real workflow management environment described in Section 6.2 working on top of the BeesyCluster middleware described in Section 6.1.

In addition to the tests conducted by the author for various service availability as demonstrated in [68], a more thorough analysis is performed in this book for a sample workflow application as shown in Figure 4.3. Using the aforementioned notation, the workflow graph can be modeled as

$$t_1 \to (t_2 \to t_3)^4 \to t_9 \to ((t_{10} \to t_{11})||t_{12}) \to t_{13}.$$

It may be suited for a variety of complex applications such as:

1. business production problems in which parts of the workflow correspond to component or service purchases, integration, and distribution as discussed in detail in Section 7.3, and

2. scientific workflow applications in which parts of the workflow denote parallel analysis of data such as images, sound clips, documents, e.g., for detection of patterns, etc. and subsequent storage, removal, notification etc.; such practical examples are discussed in detail in Section 7.2 and 7.6.

In this workflow application, tasks were assigned services as follows: t_1 up to t_8 and t_{13}—10 services with the following cost/execution times parameters: 12/8, 12/8, 10/9, 10/9, 8/10, 8/10, 7/11, 7/11, 6/12 and 6/12, t_9—one service with parameters 12/8 (as it corresponds to own integration of previous results in the aforementioned applications), t_{10} and t_{11}—services with parameters 12/8, 12/8, 10/9, 10/9, 8/10, 8/10, and t_{12}—services 7/11, 7/11, 6/12 and 6/12.

ILPHEU used the timeout of 10 seconds in the experiments and `lpsolve` [9] was used as an ILP solver. GA used 10 chromosomes per population. The divide-and-conquer solution used ILPHEU for subproblems.

Firstly, workflow execution times are shown for the algorithms for various service availability. Service availability of 100% means that all services selected before workflow execution are available at all times. Consequently, before a task is executed, the service selected previously for the task is available. Service availability of x% means that upon a check of services before a task is executed, x% of the services for the task are available. If the previously chosen service is not available, workflow rescheduling must occur. In this case, parameters of the services for already executed tasks must be taken into account and the remaining part of the workflow should be scheduled, still meeting the optimization criteria. As an example, for the workflow application shown in Figure 4.3, corresponding workflow execution times are shown in Figure 4.4. It can be seen that for lower service availability values, the total execution time of the algorithms is indeed higher due to rescheduling. The order of performance for the algorithms for this workflow application is as follows: ILPHEU, GAIN, divide-and-conquer, and the genetic algorithm.

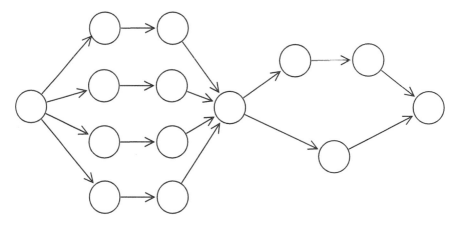

FIGURE 4.3: A testbed workflow application with 14 nodes.

However, one should keep in mind various limitations of particular algorithms that are outlined in Table 4.2. These may prevent you from using a particular algorithm for certain workflow applications. As an example, in real workflow applications, for low values of service availability, the divide-and-conquer algorithm often fails because it is not able to meet the cost constraint for partitions. Figure 4.5 shows percentage of failure of such an algorithm during a run of a workflow application with given service availability. Failure basically requires another algorithm to continue with a feasible solution. Given the parameters, the other algorithms were always able to find a valid solution. However, for tighter cost bounds, in case of failure of some services, even the others may not be able to find a valid service assignment in terms of cost requirements.

It is interesting to see how close to the imposed cost bounds particular algorithms can get. Real costs obtained by algorithms for the scheduling the aforementioned workflow application are shown in Figure 4.6. ILPHEU and

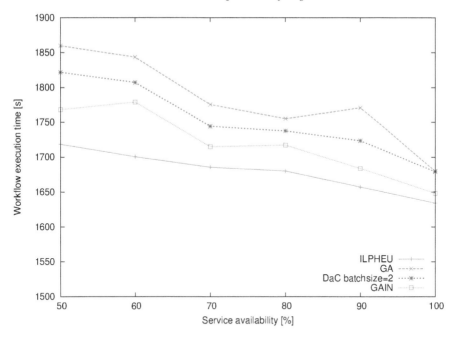

FIGURE **4.4**: Workflow execution times [s] for the workflow application shown in Figure 4.3.

the divide-and-conquer algorithms are the closest in approaching the budget limit of 1102 with GA and GAIN following. Interestingly, all in all, the total cost of used services is the lowest for GAIN while it has achieved the second best workflow execution time. In tests performed in [68], the algorithms that were the closest to the budget limit were ILPHEU, GAIN, divide-and-conquer, and genetic algorithms, although differences among the last three were small.

Figure 4.7 presents average execution times for the algorithms per initial schedule for the whole workflow graph.

Finally, recommendation of which workflow applications the algorithms are suited for is presented in Table 4.3.

In [68], the author performed a similar analysis for various other workflow applications with similar structures from 6 up to over 100 nodes. Generally, very similar relative performances were obtained. Based on the experiments performed for various workflow graphs [68], balanced and imbalanced, as well as additional experiments with workflow applications of a similar structure, the following can be concluded about the algorithms, assuming that the services selected for the execution are still available while the workflow application progresses:

1. The best and similar performance is obtained for the ILPHEU and GAIN

TABLE 4.2: Workflow scheduling algorithms—limitations.

Algorithm	Limitations
ILPHEU	It may be difficult to set a proper timeout for solving an ILP problem; too small a value may result in no solution, too high may impact the total workflow execution time.
GA	Reasonably time-consuming; it is recommended that some chromosomes start from a valid (cost wise) solution, e.g., cheapest services, otherwise it may be unable to find a feasible solution within the given time frame/after a given number of iterations.
Divide-and-conquer	May be unable to solve subproblems, especially in the following cases: partitions are small or service availability is low, which may require rescheduling for a small part of the workflow graph (see Figure 4.5).
GAIN	Its execution time depends on the number of services in the workflow graph due to potential substitutions; for similar reasons it will grow for large differences of costs between that for the initial set of services and the budget limit.

algorithms, which are able to find good solutions in reasonably short time. However, this is true on the following conditions:

- A proper timeout is selected for ILPHEU and the ILP solver. Too-short values result in the solver not being able to find a valid solution. On the other hand, too-large values can considerably increase the execution time of the algorithm without giving noticeable gain in the actual workflow execution times. Additionally, good values depend on the problem size, so in practice this may require testing prior to actual runs. For workflows with 10–40 tasks and up to around 400 services in total, the timeouts selected were 5–20 seconds.

- Not too many services in the problem for the GAIN algorithm. Otherwise, the performance drops due to the possible number of substitutions and tests within the algorithm.

2. The divide-and-conquer algorithm scores behind ILPHEU and GAIN without the ability to find good enough solutions within partitions to recover from potentially lost global solutions.

3. Not surprisingly, the genetic approach comes last due to reasonably slow convergence to good solutions. However, it is the most general algorithm that is able to consider not only linear dependencies and extensions to the model.

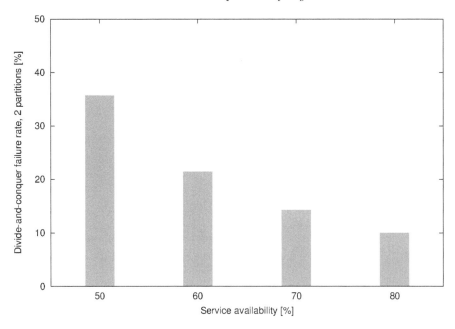

FIGURE 4.5: Failure rate of the divide-and-conquer algorithm for the workflow application shown in Figure 4.3.

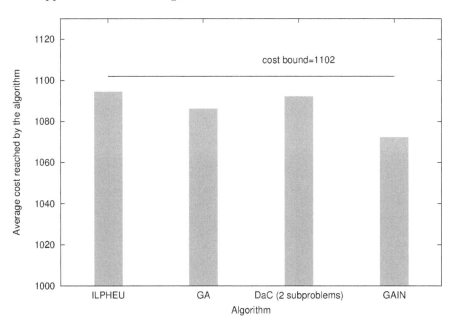

FIGURE 4.6: Average costs reached by algorithms for the workflow application shown in Figure 4.3 and cost bound 1102.

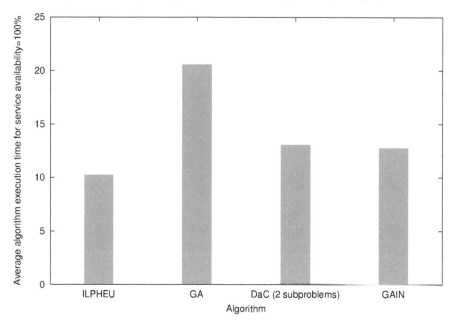

FIGURE 4.7: Average algorithm execution times for the workflow application shown in Figure 4.3, cost bound 1102, service availability=100%.

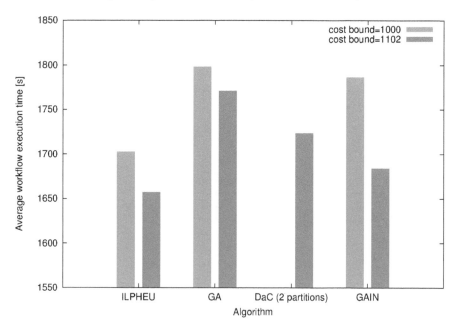

FIGURE 4.8: Workflow execution times for various algorithms and the workflow application shown in Figure 4.3, service availability=90%.

TABLE 4.3: Workflow scheduling algorithms—recommendations.

Algorithm	Recommendations
ILPHEU	Workflow applications for which a heuristic solution can be obtained within a reasonable time frame (compared to the longest workflow path considering service execution times).
GA	Workflow applications for which no parameter adjustment is possible; workflow applications with nonlinear dependencies.
Divide-and-conquer	Workflow applications for which solutions for partitions may be found in reasonable time frame when ILPHEU would not be able to find a valid solution for the whole workflow graph within such time frame.
GAIN	Workflow applications for which no preconfiguration is possible.

Paper [68], among others, also considers relative execution times of the algorithms and its contribution to the total workflow execution time as well as comparison of GA for various numbers of iterations.

4.4 Service Selection with Dynamic Data Flows and Automatic Data Parallelization

The previously proposed model assumes that each workflow task processes data of predefined data size as shown in Figure 4.9.

In some cases, though, especially if several alternative solutions can be found for input data, the latter may be modeled by parallel and alternative paths or tasks in the workflow. In this case, the task is to determine which paths the input data should take to optimize the given QoS goal. As in the formulation in Section 3.4.2, workflow tasks can have one or more services assigned to them out of which one needs to be chosen. Even if one service is assigned per task, selection of which services are executed can be performed by routing the data through particular paths with the given services. This concept, which was previously proposed by the author in [60], is extended in this chapter. Moreover, input data can be partitioned into disjoint packets that can be routed through various workflow paths in order to minimize the workflow execution time. It should be noted that in many practical problems, sizes of flows will be constrained to integer values.

Figure 4.10 depicts a workflow to generate a web album out of 100 images taken by a digital camera in a RAW format for best possible quality. It requires

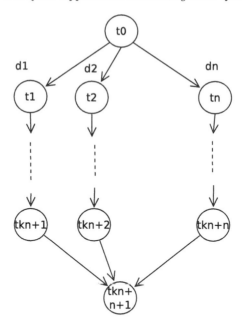

FIGURE 4.9: Data distribution

conversion from the RAW format to a JPEG and superimposing a watermark image with the copyright text. Then the resulting images would be passed to a web album generator. Each image I is of size 100 while the watermark image W is of size 2. There are two alternative methods to solve this depending on the RAW converters available (see Figure 4.10):

1. a RAW to JPEG converter, which also applies the watermark image to the input image (t_3), or

2. a RAW to PPM converter (t_4) followed by a PPM to JPEG conversion (ppmtojpeg); t_5 uses another application to apply the watermark image.

Task t_2 corresponds to just passing the watermark image to t_3 and is introduced to impose criteria on doing so. Namely, the watermark image is passed to t_3 if and only if at least one image is processed by t_3. Otherwise, there is no need to pass W to t_2 and then to t_3.

We assume that there is only one computer with one service capable of solving t_3 using a Microsoft Windows machine. On the other hand, there are two services to perform functions of both task t_4 and t_5 (like two dcraw converters installed on different machines of different speeds). This implies that t_4 and t_5 can be parallelized into functionally equivalent t_6 and t_7 correspondingly. Task t_8 converts resulting JPEG files to a web album for which there are many applications available.

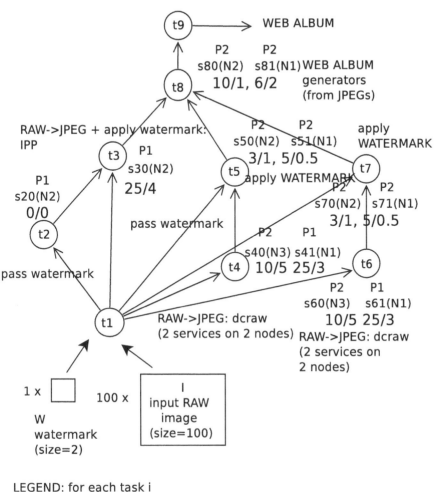

FIGURE 4.10: Exemplary workflow with data parallelization.

There is a set of services $S_i = \{s_{i1}, s_{i2}, ..., s_{i|S_i|}\}$ out of which only one must be selected to execute task t_i. These have parameters like cost (c_{ij} which denotes a cost of processing a unit of data), execution time (t_{ij} which denotes how much time the particular service spends on processing a unit of data) and are offered by particular providers. $d_{i \to j}$ denotes the size of data flowing from task t_i to t_j while d_{ij}^{in} the size of data processed by s_{ij} and d_i data processed by task t_i where $d_i = \sum_k d_{k \to i} = \sum_j d_{ij}^{in}$. It should be noted that there are specific constraints on data flows depending on what data (W or I) is passed along graph edges.

The variables to find in this problem are:

- assignment of one of the services in S_i to each task t_i,

- finding all d_{ij}s which determine data distribution between the selected services. Some links may have a fixed size of data such as $d_{1 \to 2}$, $d_{1 \to 5}$, $d_{1 \to 7}$ may be either 0 or 2 (the latter if t_3 or t_4 or t_6 are sent at least one RAW image) while $d_{1 \to 3} + d_{1 \to 4} + d_{1 \to 6} = d_3 + d_4 + d_6 = 100$, which means the set of input RAW messages may be partitioned between services for parallel execution.

We consider two alternative minimization goals where t_{workflow} is the time when the last service finishes:

1.

$$v_{MIN\_T\_C\_BOUND} = t_{\text{workflow}} \tag{4.22}$$

while adding a constraint on the total cost, i.e., $\sum d_{ij}^{in} c_{ij} \leq B$ where B is the budget (problem MIN_T_C_BOUND).

2.

$$v_{MIN\_TC} = \alpha t_{\text{workflow}} + \sum d_{ij}^{in} c_{ij} \tag{4.23}$$

(problem MIN_TC) tested for comparison with the aforementioned goal as a balance between the time and the cost ($\alpha > 0$).

Table 4.4 and Figure 4.11 present exemplary but non-optimal schedules (assignment of tasks to services and starting times) as well as data sizes assigned to particular tasks and data flows between tasks for the problem depicted in Figure 4.10.

This example can clearly benefit from both selection of services, as there are several services to execute, e.g., the web album generation from JPEGs (t_8) as well as parallelization of the RAW to JPEG conversion to shorten the execution time (t_3, $t_4 \to t_5$ and $t_6 \to t_7$). Moreover, passing the watermark image occurs if and only if the corresponding RAW converter has been activated.

Compared to the previous formulation with fixed data sizes presented in Section 4.2.1, the problem input includes the input data size as well as a formula that defines how output data from a task should be distributed among input data of the successors of the given task.

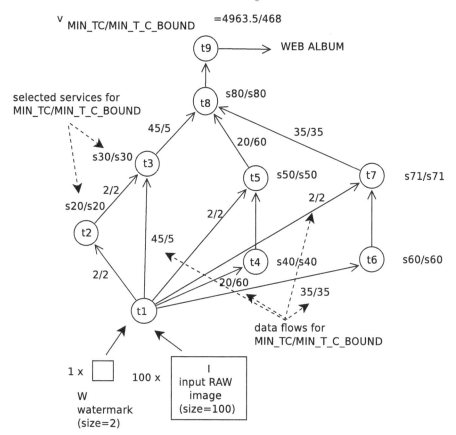

FIGURE 4.11: Non-optimal example of a workflow with selected services and data distribution.

As previously, d_{ij}^{in} and d_{ij}^{out} are assumed to be the sizes of input data and size of the data produced by service s_{ij} bound with $d_{ij}^{out} = f_{t_i}(d_{ij}^{in})$. Function f_{t_i} is related to the task, not the service, since all services executing the task must take the same input and return same output data. d_i denotes the size of data processed by task t_i. Variable d_{ijkl} denotes the size of data to be sent from service s_{ij} to service s_{kl}. Then the required constraints on these variables would include:

$$d_i = \sum_j d_{ij}^{in} \tag{4.24}$$

$$d_{kl}^{in} = \sum_{i,j:(v_i,v_k)\in E} d_{ijkl} \tag{4.25}$$

$$d_{ij}^{out} = f_{t_i}(d_{ij}^{in}). \tag{4.26}$$

Condition $d_{ij}^{in} > 0$ holds for only one selected service s_{ij} for task t_i. Assuming

TABLE 4.4: Schedule and data distribution for a non-optimal example of a workflow with selected services and data distribution.

	MIN_TC			MIN_T_C_BOUND		
v	4963.5			468		
t_{workflow}	298.75			468		
t_i	s_{ij}	t_i^{st}	d_i	s_{ij}	t_i^{st}	d_i
t_2	s_{20}	0.1	2	s_{20}	0.1	2
t_3	s_{30}	4.5	47	s_{30}	0.5	7
t_4	s_{40}	71.75	20	s_{40}	3	60
t_5	s_{50}	176.75	22	s_{50}	306	62
t_6	s_{60}	1.75	35	s_{60}	123	35
t_7	s_{71}	178.5	37	s_{71}	347.75	37
t_8	s_{80}	198.75	100	s_{80}	368	100

task t_i has successor tasks in set $Succ(t_i)$, there may be several constraints imposed on how d_{ij}^{out} is partitioned into input data of tasks in $Succ(t_i)$.

In general, among other possible configurations, this allows the following constraints:

- AF task (all output data passed to each successor):
  $t_i{:}\forall_j \forall_{k:t_k \in Succ(t_i)} d_{ij}^{\text{out}} = \sum_l d_{ijkl}$,

- DF task (output data partitioned among successors):
  $t_i{:}\ \forall_j d_{ij}^{\text{out}} = \sum_{l,k:(v_i,v_k)\in E} d_{ijkl}$,

Table 4.5 and Figure 4.12 present optimized schedules (assignment of tasks to services and starting times) as well as data sizes assigned to particular tasks and data flows between tasks for the problem depicted in Figure 4.10. Parameters $\alpha = 100$ for goal MIN_TC and $B = 1000$ for goal MIN_T_C_BOUND were used. For each edge, computed data flows $d_{MIN\_TC}/d_{MIN\_T\_C\_BOUND}$ are shown for goals MIN_TC and MIN_T_C_BOUND and the final value $v_{MIN\_TC}/v_{MIN\_T\_C\_BOUND}$ of goals 4.23 and 4.22.

In the following sections, the author presents ways to extend the algorithms to the problem with distribution of data among workflow paths.

4.4.1　Extending the Genetic Algorithm Formulation

In this case, the formulation presented in Section 4.2.2 can be extended as follows.

Each chromosome consists of the representation for the schedule shown in Listing 4.1 and additionally a distribution of d_{ij}^{out} into data of the following tasks. For AF tasks, the structure shown in Listing 4.6 can be used. For task t_i, since data is divided into all following tasks, there is an array of $|Succ(t_i)|$ $datadistribution_{ij}$ elements for this task such that $\sum_{j=0}^{|Succ(t_i)|-1} datadistribution_{ij} = 1$ and d_{ij}^{out} $datadistribution_{iy} = \sum_{l:t_k:t_k \text{ is } y\text{-th successor of } t_i \text{ out of tasks in } Succ(t_i)} d_{ijkl}$, which denotes how output data for task t_i is split into input data of following tasks.

TABLE 4.5: Schedule and data distribution for an optimized solution to exemplary workflow.

	MIN_TC			MIN_T_C_BOUND		
v	4924			443.35		
t_{workflow}	297.2			443.35		
t_i	s_{ij}	t_i^{st}	d_i	s_{ij}	t_i^{st}	d_i
t_2	s_{20}	0.1	2	s_{20}	0.1	2
t_3	s_{30}	5.6	47	s_{30}	278.55	6
t_4	s_{40}	65.6	23	s_{40}	112.75	41
t_5	s_{51}	183.55	25	s_{51}	319.8	43
t_6	s_{60}	1.6	32	s_{60}	2.75	55
t_7	s_{70}	163.2	34	s_{70}	286.35	57
t_8	s_{80}	197.2	100	s_{80}	343.35	100

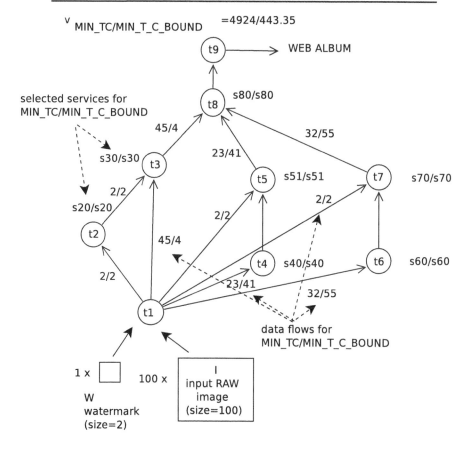

FIGURE 4.12: Optimized solution to exemplary workflow.

Listing 4.6: Chromosome in the genetic algorithm.

```
typedef struct {
    int *solution; // an array –
    // the first |T| elements denote service indexes for tasks
        i.e., solution[i]=service index for tᵢ
    // next |T| elements denote task ids – and the order of
        execution of particular tasks
    float **datadistribution; // an array of |T| arrays each
        of |Succ(tᵢ)| length
    // that denotes how output of tᵢ is partioned among |Succ(tᵢ)|
        successors
    double score;
} t_chromosome;
```

The crossover and mutation operators for the schedule shown in Listing 4.2 are to be appended with those for the data flow distribution shown in Listings 4.7 and 4.8.

Listing 4.7: GA's crossover for data distribution.

```
probability=(float)random()/RAND_MAX;
if (probability<CROSS_THRESHOLD2) {
   for(k=0;k++;k<|T|)
      new_chromosomes[bestcount+j].datadistribution[k]=
         temp_chromosome_1.datadistribution[k];
} else if (probability<2*CROSS_THRESHOLD2) {
   for(k=0;k++;k<|T|)
      new_chromosomes[bestcount+j].datadistribution[k]=
         temp_chromosome_2.datadistribution[k];
} else
   for(k=0;k++;k<|T|)
      SetRndDistr(new_chromosomes[bestcount+j].
         datadistribution[k]);
```

Listing 4.8: GA's additional mutation for data distribution.

```
for(k=0;k++;k<|T|) {
   probability=(float)random()/RAND_MAX;
   if (probability<MUT_THRESHOLD3)
      SetRndDistr(new_chromosomes[t_index].datadistribution[k
         ]);
}
```

Values of CROSS_THRESHOLD2=0.35 and MUT_THRESHOLD3=0.05 were used for all experiments. Function **SetRndDistr** assigns the flow to only one successor with probability 0.5 or divides it into more following tasks with smaller probabilities. This ensures that the algorithm can both direct the flow to a small and a large number of following tasks.

This version can be adjusted to use only integer data sizes. This just requires rounding d_{ijkl} values which result from $datadistribution_{ij}$s to integer numbers such that constraints are satisfied and successive evaluation of the chromosome uses integers.

4.4.2 Extending the Integer Linear Formulation

Extending the original problem stated in Section 4.2.1 to computing data distribution apart from service selection and scheduling is straightforward. The input data includes the data defined in Section 4.2 with the following modifications:

1. the input data size to the whole workflow is defined a priori;

2. formulas that define how output data size from each task relates to input data of successive tasks; this is stated by formulas 4.12 and 4.13;

3. individual d_is are to be determined, although may be predefined a priori for selected workflow tasks.

The optimization goal remains the same and the ILP engine is able to solve the problem assuming linear dependencies.

Furthermore, it should be noted that d_is may be either real numbers or imposed to be integers. In the latter case, the problem becomes much harder to solve. It may require heuristic approaches. Some ready-to-use ILP solvers such as lp_solve allow searching for suboptimal solutions within the predefined timeframe. Indeed, option -timeout <timeout> allows you to set the requested timeout in seconds. As noted before, though, for large workflow sizes, the solver might not be able to return a solution and may require extension of the timeout.

4.4.3 Mixing Two Algorithms for Service Selection and Data Distribution

The formulation of the problem using mixed integer linear programming allows us to use known methods to solve the problem but makes this approach impractical for large numbers of tasks and specifically a large number of services per task and a small number of processors. While in the classical service selection problem it is necessary to select services per task (integer variables), in the formulation above there can be numerous constraints and thus integer variables on the order of services and a large number of variables representing data flows between services of connected tasks. Especially, increasing the number of tasks with a small number of processors increases the number of order constraints considerably while increasing the average service count per node increases the number of data flow variables considerably.

To overcome these limitations, the author proposed to use linear programming of the problem with only real variables (i.e., data flows only) while adopting a dedicated algorithm for both selection of services and ordering [60]. For the latter the genetic algorithm can be used. Thus the algorithm would operate as shown in Listing 4.9. The main loop corresponds to steps of the genetic algorithm in which a chromosome represents both the selection of services and their ordering.

The data flow sub-problem can be then solved reasonably quickly as several constraints can be removed or simplified, having selected services for particular tasks:

1. only d_{ijkl} for the selected services s_{ij} and s_{kl} must be considered,

2. b_{ij} can be replaced with 1 and d_{ij}^{in} with d_i for the selected service s_{ij} with $\forall_{v \neq j} d_{iv}^{in} = 0$, $b_{iv} = 0$,

which reduces the number of variables considerably and makes it possible to solve the problem with data flows and ordering for a large number of tasks and services per task.

Listing 4.9: Mixed genetic and linear programming algorithm.

```
// now initialize the solutions
for (each solution) {
    select service for each task randomly;
    generate task permutation;
5 }
// iterate over the genetic algorithm steps
for(t=0;t<genalgstepcount;t++) {
    // first copy solutions to solutionsold
    remember solutions in solutionsold;
10  // now evaluate all solutions
    for(i=0;i<solutioncount;i++)
        // evaluate solution using the selected
        // services and their order using LP
        evals[i]=eval(...);

15
    // now select the minimum of evals
    mineval=min_i(evals[i]);

    copy the best solution to solutionold[0];
20  copy to solutionsold[1+] only solutions:
    solution[i] : eval[i] <= 1.05mineval;

    change each eval to maxeval*100/eval[i];
    // lower eval corresponds to a larger value;

25
    // select candidates for final solutions
    for(i=1;i<solutioncount;i++) {
```

```
        solution1 , solution2=out of solutionsold
        for crossing with probability
30      corresponding to eval of the solution;
        solutions[i]=crossing(solution1 , solution2);
    }

    // now apply mutation except solution[0]
35  for(i=1;i<solutioncount ; i++)
        mutation(solutions[i]);
}
```

It is possible to couple a different algorithm for service selection with computing data flows using linear programming. For instance, the GAIN-like algorithm presented in Section 4.2.4 could be used assuming function `EvaluateWorkflow()` includes evaluation of the workflow with computation of data flows. Unfortunately, the complexity of this approach is too large given the already too large computing time of the original GAIN2 algorithm. Thus, GAIN1 could be used with periodic checking for better data flows.

4.5 Modeling Additional Dependencies or Constraints

The model allows us to specify several other specific dependencies or constraints that might be useful in specific workflow applications or contexts. This section proposes how to model these issues.

4.5.1 Discounts

Services or packets of services for various workflow tasks can be offered at a discount price when purchased together. Alternatively, a scenario in which such services are not selected may be considered as one that requires a penalty. For instance, for services s_{ij} and s_{kl} this corresponds to an additional penalty if $b_{ij} + b_{kl} < 2$. Consequently, a penalty could be added if $2 - b_{ij} - b_{kl} > 0$ which requires just adding $p(2 - b_{ij} - b_{kl})$ to a minimization goal $(p > 0)$.

In real applications, special offers may be valid for purchasing several products or services, either from one or many providers. The same can be applied for workflow applications and services needed to accomplish particular workflow tasks. For a packet of n services, the above formula can be extended to n services as: $n - b_{ij} - b_{kl} - \dots$. Alternatively, for two tasks, the following formula can be added to the overall cost of selected services:

$$p|b_{ij} - b_{kl}| \tag{4.27}$$

where $p > 0$ is a weight. If either both or none of the services are selected, the penalty is 0. This can be extended to include combinations of other services as well.

4.5.2   Synchronization Tasks

It is possible to introduce control tasks into a workflow application, aimed at just synchronization of selected tasks. In this case, an additional task is added with execution time set to 0 and incoming and outgoing data flows equal to 0 (Figure 4.13). Constraint 4.17 will assure synchronization of the preceding tasks.

4.5.3   Associated Tasks

In the model that requires optimization of data distribution among parallel workflow paths (Section 4.4) it is possible to define association of tasks, i.e., task t_i will be sent data by its predecessors if and only if another task t_k is executed. For instance, in Figure 4.10 $d_2 = 2$ if and only if $d_3 > 0$. If I is not sent to t_3, there is no need to send W to t_2.

$$d_{i \to j} = \begin{cases} d_i & \text{if } d_k > 0 \text{ since } t_j \text{ needs data only if } t_k \text{ is executed} \\ 0 & \text{otherwise} \end{cases}$$

The following constraints can be added:
$$d_{i \to j} = \sum_{(v_i, v_j) \in E} d_{iajb}$$

$$\left| d_{i \to j} - \frac{d_i}{2} \right| \geq \frac{d_i}{2}$$

to force $d_{i \to j}$ to be either 0 or d_i. Then:

$$d_{i \to j} \geq d_k / D_k$$

for a sufficiently large constant D_k to activate t_i if t_k is active.

4.5.4   Integer Data Sizes

The model presented in Section 4.2 assumes data sizes are known a priori. On the other hand, the model with automatic data distribution among workflow paths presented in Section 4.4 allows forcing $d_i \in Z$ $d_{ijkl} \in Z$, which makes partitioning into integer values only. Consequently, it will make the problem more difficult to solve.

4.5.5   Data Sizes Known at Runtime Only

In the models assumed so far, data sizes processed by services selected for particular tasks were either known a priori or determined during an opti-

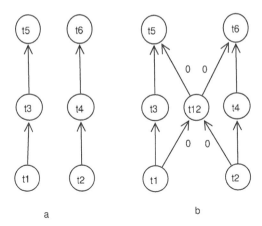

FIGURE 4.13: Task synchronization.

mization algorithm. In the latter case it allows data parallelization and consequently high throughput in a workflow.

In some cases, however, the size of output data from a service may only be determined and generated at runtime and cannot be predicted in advance. Even such a scenario can be handled though. Similar to unavailability or failure of a service, a rescheduling request would be issued to an execution engine. The latter would launch an optimization algorithm with data flows to following tasks reset to new values. It is probable that new services would need to be selected for new data sizes if the same constraints are to be met again.

4.5.6 Data Servers or Repositories

In the proposed model, it is assumed that output data from a service is sent as input data to a service assigned to a following task or services assigned to following tasks. In the practical realization of the copying phase the following factors need to be taken into consideration:

1. For direct transfer of output data to service assigned to a following task, the current service would need to possess information about the workflow. Consequently, several services assigned to parallel tasks would manage data transfers independently. In case of failures, recovery from failure would require global synchronization in order to reselect services meeting global criteria.

2. Workflow execution is managed by an execution engine. The engine

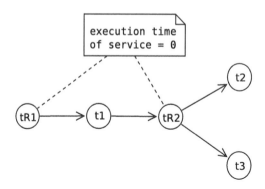

FIGURE 4.14: Communication using data repositories.

could order a data transfer from the location of the current service to a location of a service assigned to a following task. Alternatively, data transfers could be executed via the execution engine itself. In the latter case, communication parameters would need to consider these links.

3. It is possible that a service will need to fetch data from a dedicated repository and save it to a dedicated repository before a service assigned to a following task reads it. Such a scenario can be modeled as shown in Figure 4.14 where special tasks were introduced corresponding to routing the data through the given repository. Such a task is assigned one service with execution time set to 0. Instead, communication costs will result in the actual overhead for such an operation.

4.6   Checkpointing in Workflow Optimization

As outlined in Section 2.2.6, there are a few reasons for the need of rescheduling of a workflow application at runtime. Those referred to changes of service parameters, unavailability, or new services becoming available before a service for an upcoming task is to be launched.

However, in some circumstances it is possible that parameters of an already running service may change. This might be due to changes of the environment in which the service is running. A workflow execution engine knows an expected execution time for service s_{ij} as t_{ij}^{exec}. If the service has not returned results within the expected time frame and an additional timeout, the engine may declare that the service has failed. In this case, without additional knowledge about the node on which the service started N_{ij}, the engine would need to start a workflow rescheduling process as if the service was unavailable. However, the service might still be running but may actually be slowed down by additional load on node N_{ij}. This scenario is shown in Figure 4.15.

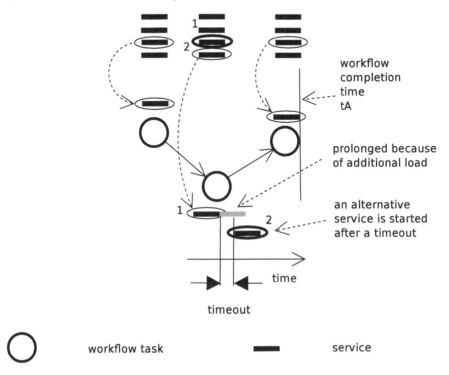

FIGURE 4.15: Starting an alternative service after a timeout from an expected service termination time.

It is possible to optimize such a scenario by getting more knowledge about the environment in which the service runs. In [57] the author presented a solution in which additional services are launched for monitoring loads and checkpointing and restarting an application behind a service to another node. This allows detection of load caused by other processes and subsequent checkpointing an application behind a service on this node and restarting on a different part of the cluster.

This concept can be implemented within the workflow optimization model considered in this book as follows. Some services and applications running behind are good candidates for checkpointing as opposed to others. Those that run in a cluster environment with possibly many nodes sharing a file system and libraries may be easily moved to another node or part of the cluster. In other cases, applications behind services will not be moved because:

1. providers do not allow executables to be downloaded,

2. the application requires a specific environment or other software only available on this node.

A service may launch an application in parallel with a process monitoring load. The latter, upon detecting a considerable change in the load on the node, would locally investigate potential nodes for migration and select one, e.g., with a lowest load for migration. In a cluster-oriented environment, checkpoint files saved on one node might be easily accessible from another through a shared file system. Two options need to be considered at this point:

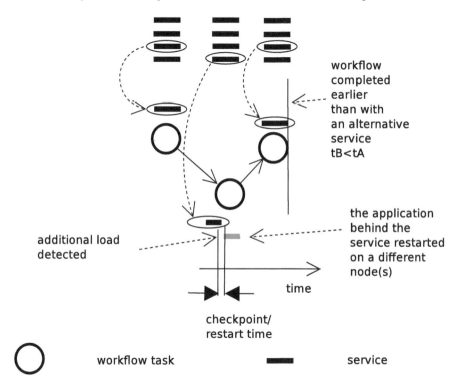

FIGURE 4.16: Checkpointing an application behind a service and restarting on a different node.

1. If it is possible to checkpoint the application behind service $s_{c\ sel(c)}$ and restart computations on a different node such that the total execution time including the checkpoint/restart remains within the aforementioned timeout of the execution engine, computations may be continued with only a slight delay (Figure 4.16).

2. Otherwise, workflow rescheduling is launched, which is more costly due to workflow-level optimization. In this case, additional services might be considered that correspond to the partial results from the checkpoint and continuation on other nodes with possibly various speeds.

Chapter 5

Dynamic Data Management in Workflow Execution

In today's information systems, big data and its processing are becoming more and more important. This is possible for two reasons:

1. Widespread availability of various data streams such as:

   - data from sensors even in widespread mobile devices; such data can be used for various applications such as collection of measurements during physical activities [44]; measurements of air pollution using mobile phones, thanks to the Visibility app for Android smartphones developed at University of Southern California [113, 173]; noise monitoring in the city [149]; recording and passing latest information to news portals;

   - recording video such as from RGB-D cameras for further analysis of, e.g., verification of correctness of movement during rehabilitation etc.;

   - data of large sizes from radio telescopes [160], [181].

2. The increasing computational power of devices as discussed in Section 2.1.1 at the level of multicore CPUs, accelerators, and clusters.

The aforementioned examples will often be modeled and executed as workflow applications. In this context, constraints on data sizes may appear in the following points (Figure 5.1):

1. services for either data acquisition or processing,

2. the management layer supervising the workflow execution.

This chapter will investigate data processing in workflow applications considering data parallelization, constraints on buffers and storage, and batch and streaming processing. Original solutions to these problems will be provided along with experiments of real workflow application examples executed in distributed environments.

5.1   Model Considering Data Constraints and Management

In order to accommodate various ways of data management and data constraints, the notation includes variables concerning data, in addition to the model considered previously in Section 4.2. This follows the work of the author from [64]. Relevant constraints are marked in Figure 5.1.

Taking into account the storage parameters and the constraints, workflow application scheduling and corresponding optimization algorithms defined in Section 4.2 can be used for the extended model. It is the calculation of the workflow execution time that needs to be modified and consider the constraints. This part is presented in Figure 5.2 extending the approach proposed in [64]. The algorithm uses the notation introduced in Section 3.4.5 and extends Figure 3.10 with data management using the following notation:

- ds_{ij}^{max}—the maximum storage capacity of the machine on which service s_{ij} is installed and where it runs; it is related to hardware resources; potentially various configurations could be considered that would be associated with various corresponding service costs.

- d_{input}—size of the input data that is passed to the workflow.

- $WEC(t)$—the maximum size of data that can be held at time t in the storage space used by the workflow execution engine for copying output data of services to services assigned to following tasks.

- $SDP_{i;k}$—the size of a data packet that is sent between tasks t_i and t_k; actual possible values may be determined by a particular application.

- $PARDS_{dl;i}$—the number of parallel streams used for passing data from initial data location dl to the service assigned to task t_i of the workflow application.

- $PARDS_{i;k}$—the number of parallel streams used for passing data between services selected for execution of tasks t_i and t_k such that $(t_i, t_k) \in E$.

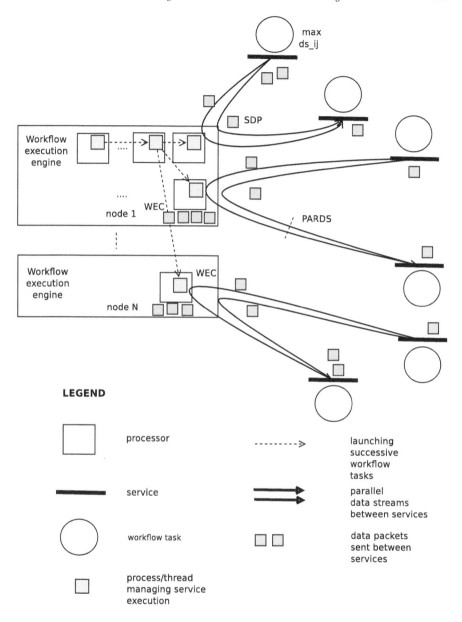

FIGURE 5.1: Multithreaded model for management of data during workflow execution.

- $t_{\text{workflow}} \in R$—the wall time of workflow execution, i.e., in case of the streaming mode it is marked by the time when the last service finishes processing the last chunk of data.

- $B_{WEC} \in R$—budget allocated for the cost of storage used by the workflow execution system; it may be up to the scheduling algorithm what amount is allocated to B_{WEC}.

Compared to [64], Figure 5.2 considers workflow rescheduling as well as non-streaming and streaming modes for tasks.

5.2   Multithreaded Implementation of Data Handling

This section proposes a multithreaded implementation of data handling in workflow execution. The following thread classes are distinguished for management of parallelization:

- Threads managing execution of particular services (Listing 5.1), i.e., invocation of the service and passing input data as well as waiting for results and passing output data to threads responsible for execution of subsequent workflow tasks. The main steps for execution of a service for a workflow task are as follows:

  1. Fetching the workflow instance identifier, the identifier of the task to be executed, and the identifier of the user logged in and launching the task.

  2. Fetching a reference for a solver for the workflow that has the information about all workflow tasks and services chosen to execute the tasks.

  3. Launching the previously selected service for the task; this step checks the user credentials against privileges granted for the service.

  4. Reselection of a service after the previous one has failed or not finished within a predefined time frame.

  5. Updating the cost of the workflow based on the executed service.

  6. Reselection of a service for the following workflow task if the previously selected one is not available or new services have appeared.

  7. If streaming has been enabled for the following task, then create DataCopier threads and assign output data files for passing data to a subsequent task; this automatically invokes a new thread for handling the successor; each instance of a DataCopier may have data assigned up to a predefined limit in terms of the size.

  8. If streaming has not been enabled, copy files to the location of the following service (without starting processing).

  9. Wait for completion of the started DataCopier threads.

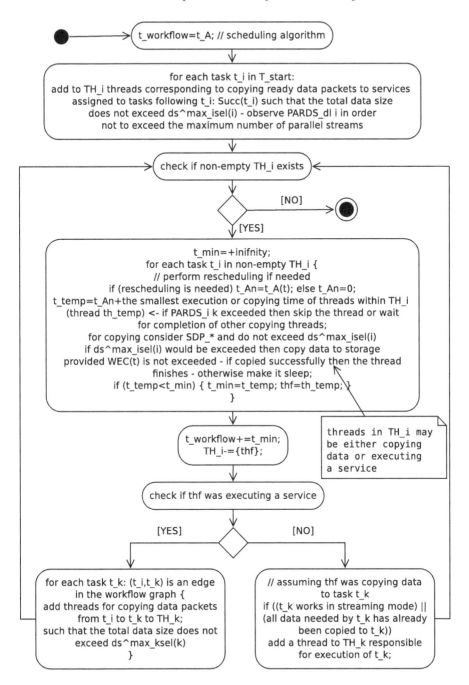

FIGURE 5.2: Evaluation of the workflow execution time—extending Figure 3.10 and [64] with data constraints.

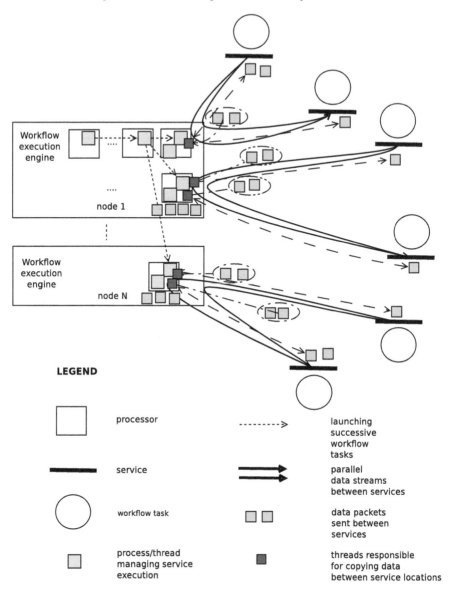

FIGURE 5.3: Multithreaded model for management of data during workflow execution with threads for copying data.

10. Remove the files used by the current thread:
    - in the batch mode this thread already copied the data to the location of the following service,
    - in the streaming mode with data caching DataCopier threads

created new threads after the data had been copied to a cache; consequently the data can be removed from the previous location.

11. Unlock execution of the current node and notify the DataCopier threads waiting for initiation of new instances of it.

12. Initiate execution of the following node if the following task does not have streaming enabled.

- Threads distinguished for management of data transfers—in this case, the threads can group data packets before sending; additionally it is possible to pack many small files into an archive and unpack after the transfer has been completed. There is a trade-off between the increased cost of sending multiple small files and the overhead for packing and unpacking the files. The main steps for such a thread are as follows:

  1. When the successor allows (no other DataCopier instance is inside the critical section), execute the following.

  2. There is data of a certain size assigned to the thread; the thread will be copying data packets with size determined by minimum of this size, the maximum storage space left on the location of the service assigned to the following task, and the maximum size of a data packet to be sent between services.

  3. There may be more DataCopier threads trying to copy data (there may also be more predecessors to the following task); if there is not enough space the thread will not be able to copy.

  4. If the data was not copied, then retrieve the data to an intermediate cache; if the cache is full, wait until the data can be copied and continue; subsequently update the size of data occupied in the cache; access to the cache is synchronized.

  5. Create and start a thread for copying the data to the location of the following service; this allows to separate initiation of the service for the following task from the previous service; this way the output data can be deleted from the location of the previous service and potentially allow start of processing of new data packets by the previous service.

  6. Within the thread, data is copied and processing of the service is started.

Listing 5.1: Pseudocode of a thread responsible for execution of workflow task.

```
public class TaskHandler implements Runnable {
  (...)
    public void run() {
```

```
       (...)
       //get a reference to the workflow solver for this
            workflow instance
       WorkflowMappingSolver  workflowSolver=
            WorkflowMappingSolverPool.getInstance().
            GetMappingSolver(workflowInstanceId);
8

       invoke the service selected by the solver for
       the node  — workflow node for each of the input
       files available at the moment;

13     reselect a service if the current one has failed a
            predefined
       number of times and run the service until succeeded
       or no more alternative services available;

       add the cost for the current node synchronizing on the
            solver instance;
18
       if (service for the following node not available ||
            new services appeared) {
         synchronize and select a new service for that node;
       }
       if (streaming enabled for the task) {
23       create DataCopier objects and fill them with
         the output data to be sent to the following nodes;
         // considering the output data from this service for
              other data parts
         // were already sent and data sizes to be sent
         // to particular tasks as indicated by
              workflowSolver
28       if (successor.getStreaming()==true) {
           start the DataCopiers (threads copy data) — the
                threads will initiate
           services for the following node after copying;
         } else
         copy files to the location of the following service;
33     } else {
         if this is the last portion of input data for the
              service
         distribute the output data to the following nodes;
       }

38     for(int k=0;k<dataCopierIndex;k++)
         dataCopiers[k].join();

       remove the files (input/output) used by this thread;
       decrease the number of files used by this task;
```

```
43    unlock execution of this node and wake up threads
      that may have been waiting for copying new data
      to the location of this service;

48    if this is the last portion of input data for the
          service
        initiate successors for which successor.getStreaming
            ()==false; // the data // should already be there
    }
  }
```

5.3 Data Streaming

The proposed solution allows for selection of one out of two working modes for each workflow task:

1. *Batch processing*—all input data for a task must be collected from the previous task(s) before the service selected for the given task is launched. By design, certain tasks must be executed this way. As examples:

   - Business—a certain number of offers must be collected before selection (the service) is invoked.

   - Multimedia—generation of a Web album requires all input images to be present.

   - Scientific—integration of data into a report and statistical analysis may require all previously planned results.

2. *Streaming*—the service selected for a given task may be invoked several times during workflow execution on data packets that are given as input. It is assumed that each data packet is processed independently by service instances. Upon termination of processing, output data is sent to the services selected for the following services while new data packets can be accepted as input. Data packets can also be grouped into chunks and given as input. Real-life examples of a workflow task using the streaming mode include:

   - Business—searching for offers where input to a service is, e.g., the address of a company (email) to check if there is an offer from them.

   - Multimedia—processing individual images through a given filter. Such a task can be a stage in a pipeline processed in the streaming mode [71].

- Scientific—generation of various tests on the same or various input data on several resources.

It should be noted that a single workflow application can accommodate tasks working in various modes. The aforementioned examples include particular tasks that might appear in a single workflow application in the given category.

5.4 Data Sizes and Storage Constraints

5.4.1 Impact of the Message Size

The size of the message sent between services has an impact on the workflow execution time due to the following trade-off:

- a small message size results in the need for sending many messages between services to process all input files. Each send operation requires overhead (startup time), which in total may be too large due to the number of messages,

- a large message causes the subsequent service to wait longer for data to process, which delays the start of processing.

For this reason, it is necessary to find the optimal data size to minimize the overall workflow execution time. This issue is illustrated by results shown in Figure 5.5 for the workflow application depicted in Figure 5.4. The settings of this experiment were as follows: 80 input files (data chunks) of 12 MB each, $ds_{ij}^{\max} = SDP$ files, $PARDS = \{1, 2\}$, streaming enabled. SDP is represented on the X axis. The goal was to minimize the workflow execution time, and the same costs were considered for services with an infinite bound (budget). The workflow application has two parallel paths, each with three tasks that implement the following three operations on input images:

1. conversion from the RAW format to a TIFF using the dcraw [50, 184] converter (tasks t_2 and t_5),

2. normalization of the image using ImageMagick [200] (tasks t_3 and t_6),

3. resizing, sharpening, and conversion to JPEG using ImageMagick (tasks t_4 and t_7)

with task t_1 for data partitioning and t_8 for final integration of the JPEGs.

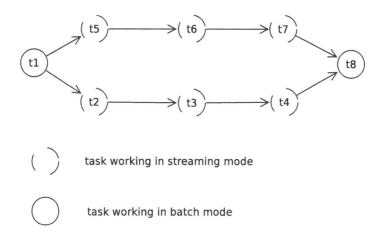

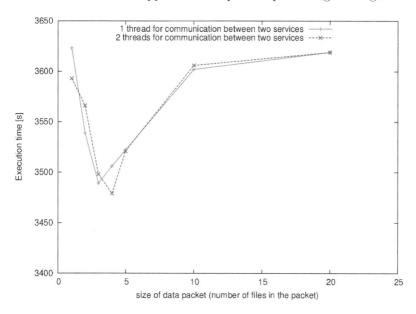

FIGURE 5.4: Workflow application for parallel processing of images.

FIGURE 5.5: Impact of the data packet size on the workflow execution time.

5.4.2 Storage Size of the Workflow Execution System

A workflow execution system always has storage constraints, whether it is centralized, as further discussed in Section 6.2.2.1, or distributed, as proposed in Section 6.2.2.2. The constraints, depending on the number of streams in a workflow application and sizes of data passed, may have serious impact on the workflow execution time. Namely, if data is passed between services through the execution system, it will be staged if the maximum capacity of the execution system has been reached by previously sent data. As soon as the buffers have been released to the point that the next data packet can be stored, the latter is copied and can be deleted from the location of the previous service. As an example, for the workflow application shown in Figure 5.6, Figure 5.7 presents workflow execution times for input data and parameters: $d_{\text{input}} = 80 \cdot 12MB$, $PARDS_* = 8$, $ds_*^{\max} = 10$, $SDP_* = 10$, with streaming enabled, while Figure 5.8 presents workflow execution times for data set: $d_{\text{input}} = 80 \cdot 100MB$, $PARDS_* = 8$, $ds_*^{\max} = 10$, $SDP_* = 1$.

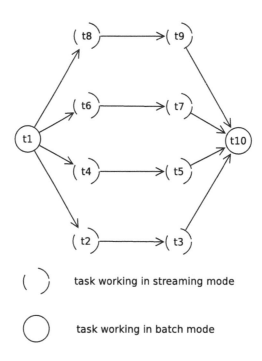

FIGURE 5.6: Workflow application for testing the impact of the storage size of the workflow execution system on the workflow execution time.

Assuming that storage of the execution engine entails corresponding costs, the initial budget B needs to be allocated not only to the costs of services (that might include the storage of input data when running a service) but also to the costs of the workflow execution system. This problem and a solution are considered by the author in [64].

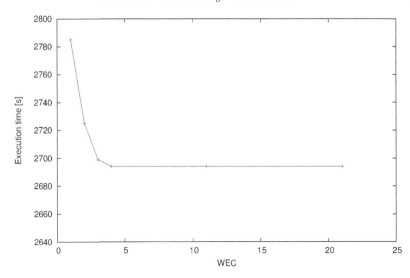

FIGURE 5.7: Impact of the storage size of the workflow execution system on the workflow execution time, $d_{\text{input}} = 80 \cdot 12MB$, $PARDS_* = 8$, $ds_*^{\max} = 10$, $SDP_* = 10$.

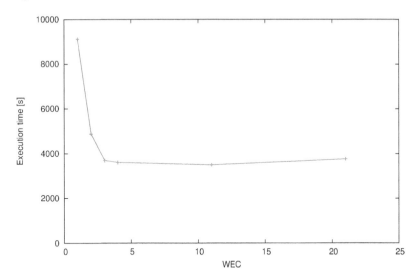

FIGURE 5.8: Impact of the storage size of the workflow execution system on the workflow execution time, $d_{\text{input}} = 80 \cdot 100MB$, $PARDS_* = 8$, $ds_*^{\max} = 10$, $SDP_* = 1$.

5.4.3   Storage Constraints for Service Nodes

Similar to the constraints of the workflow execution system, nodes on which services are installed may have storage constraints. Two possibilities can be distinguished in this case:

- *Batch processing mode*—in this case the service can run only one instance at a time. All input data must be ready and available in order to start the service.

- *Running service instances in parallel*—invoking the service in parallel with various input data creates potentially many instances of the service executing in parallel on the server. In such a case, if the service uses shared resources on the server such as disk space, proper synchronization mechanisms such as locks, conditional variables, and notifications such as those proposed in [57] might be required. Additionally, running several instances in parallel may require large intermediate storage in the location of the service. If there is a limit on this storage, running another instance may be stopped until space is available. This may impact the workflow execution time because the space which was used for output results in the location of the previous service could not be released. Consequently, running the previous service using another data chunk will be stopped.

5.4.4   Data Caching for Increased Throughput and Improvement of Workflow Execution Time

One of the solutions available to mitigate the issues with storage constraints discussed above—both of the workflow execution engine and the service nodes—can be to provide the system with additional caches to store output data of the previous services until data can be sent further, either copied to the location of the execution engine or to the service itself. The advantage of this solution is to release the storage location of the previous service so that processing of the previous service can follow while previous instances of the following service are finishing.

In [64] the author presented tests showing the impact of data caching on the workflow execution time for a workflow application with a fork structure. As an extension to that result, Figure 5.9 presents results for the workflow shown in Figure 5.6 and with larger input. Data and parameters for this test are as follows: $d_{\text{input}} = 80 \cdot 12MB$, $PARDS_* = 2$, $ds_*^{\max} = 2$, $WEC(t) = +\infty$. Services in all paths have the same execution times.

5.4.5   Runtime Workflow Data Footprint

The proposed approach, especially the proposed multithreaded implementation shown in Section 5.2, can be easily extended to consider the total run-

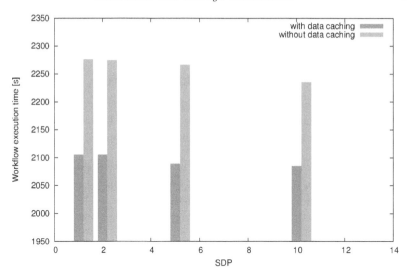

FIGURE 5.9: Workflow execution time without and with intermediate data caching.

time workflow data footprint and an upper bound on it. The total workflow footprint of workflow application A at time t will be equal to:

$$DF(A,t) = \sum_i d_i(t) \qquad (5.1)$$

where $d_i(t)$ is the total data size of all instances of service s_{ij} chosen to execute task t_i. In fact, in case of failures it is possible that data packets processed later might be processed by another, functionally equivalent service. Then, assuming that for each task t_i there are data packets $d_i^p(t)$ processed at particular moment t (might be in various stages of processing) then:

$$d_i(t) = \sum_p d_i^p(t). \qquad (5.2)$$

The execution engine might control the runtime data footprint by not allowing more data packets into the workflow application until the data footprint with the new data fits below a predefined threshold. Then, the workflow scheduling problem defined in Section 3.4.2, might consider another constraint, i.e.,

$$\forall_t DF(A,t) \le DSMAX. \qquad (5.3)$$

Obviously, setting tight DSMAX values may impact the workflow execution time in the same way as the storage constraints of the services and the workflow execution engine. On the other hand, in resource-limited environments, it may allow running many instances of workflow applications at the same time, either for one or more users.

5.4.6   Cost of Storage in Workflow Execution

Work [64] considers the cost of storage of files in the workflow execution system and optimization of allocation of a part of the initial budget B to the storage of the execution system B_{WEC} and the remaining part to the cost of the services.

That model can be further extended as follows. By default in the model presented in Section 3.4.2, cost c_{ij} is associated with service s_{ij}. Taking the storage size available for the service, various contracts might be available for the service consumer. Namely, various cost/storage options might be available. The provider might distinguish various (c_{ij}, ds_{ij}^{max}) contracts that might affect corresponding t_{ij} values. For instance, based on the proposed workflow execution algorithm using data streaming, the provider might simulate the execution of a workflow application composed of just one service for various d_{input} values as well as various SDP_* values to show how this affects the execution time of a service. Then, the provider obtains final (c_{ij}, t_{ij}) contracts that they might make available to the user. Those contracts can be seen as separate, functionally equivalent services for the workflow application scheduling algorithm.

5.5   Service Interfaces and Data Interchange

The model proposed in Section 3.4.2 assumes that each task has a set of services assigned to it that are functionally equivalent. At the same time, each of the services is assumed to be able to read output data from the previous service as well as produce output data that will be compatible with services assigned to following tasks. This raises the following issues, i.e., how to:

1. find services capable of executing the given function defined for the task, and

2. ensure that data formats will be compliant.

Solutions to these issues were proposed by the author et al. in works [76] and [81], respectively. The following paragraphs propose extensions and further ideas of how those can be incorporated into workflow management.

Matching cooperating services can be based on the functional specifications of the service, namely inputs, outputs, precondition, effects, and the function performed by the service.

Firstly, based on the idea proposed in [198], this description can be used for finding services for the given workflow task. Each task in a workflow can

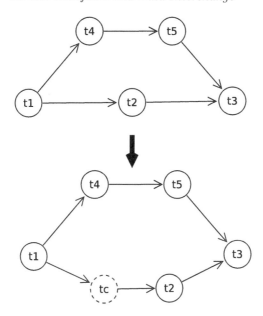

FIGURE 5.10: Extension of a workflow model with a task for conversion service.

be described by values assigned to the aforementioned elements, i.e., inputs, outputs, precondition, effects and the function performed. If we assume that existing services are also described in such a way, as proposed by the author in Figure 3.1, then assignment of best services matching task requirements can be found. Namely, it is possible to compute the distance between the description of the service and the task specification based on the type of relation between the two (no relation, inheritance, and the difference between the level numbers in the inheritance tree) which can be expressed as a numerical score.

Secondly, as proposed by the author et al. in [81], even if descriptions of services are not known in advance or are only partially known, it is possible to adopt a learning technique to complete the descriptions using the following scheme:

1. Select a service out of the set of services with known interfaces with certain probability p and other services with probability $1 - p$.

2. Based on the result of the service with, e.g., an unknown input description and the predecessor service with a known output complete the description of the unknown service. This can be done automatically based on whether the unknown service returns valid results. The latter requires a validator to determine correctness of results unless any result other than predefined error codes is assumed to be correct.

Thirdly, the previously proposed solutions can be further extended for even more automatic service composition. If the search engine finds a service that functionally matches the task requirement but, e.g., one of the formats (input, output) is not fully compatible with the task specifications, it can expand searching for services for a task to searching for, e.g., two services, one for performing the task and the other for conversion of formats. Both the execution time and the cost will be affected by inclusion of the conversion service. It is still possible that matches with a conversion service will surpass ones without and thus offer better solutions to the client. This scheme is depicted in Figure 5.10 where tc denotes the task to which a conversion service is assigned.

Chapter 6

Workflow Management and Execution Using BeesyCluster

6.1   BeesyCluster: A Middleware for Service-Based Distributed Systems

6.1.1   Middleware Requirements

Before distributed services can be integrated into workflow applications, there needs to be a middleware available that allows definition and deployment of services to which various types of software can be mapped as proposed in Section 3.3. For cooperation of services, the middleware would need to support the key functions outlined in Section 3.4.1. The following functions allow implementation of the aforementioned requirements within the middleware:

1. Publishing various types of services (especially scientific and business ones) from servers managed by users very easily. It is possible to define access rights to others on a per-user or per-group policy. It is desired that the provider publishes a resource (application-sequential or parallel, files, etc.) from their own accounts which allows others invoking such a service on their accounts but alternatively allows us to publish installation versions for others to download and use services on their own accounts.

2. Easy and instant attachment of new servers/clusters from which new services can be published. We assume that users who wish to publish, e.g., applications installed on regular operating system accounts would not be required to install any piece of software nor execute a complex procedure to make such an application available as a service.

3. Storing information about published services in a searchable registry.

4. Hiding details of various types of middleware that act as proxies in accessing the services. This applies to the most widespread middleware types identified in Chapter 3.1. This way, invocation of a service in the middleware should be transparent irrespective of the underlying middleware on the actual resource.

5. It should be possible for the services to use drivers to control external devices connected to the particular server, e.g., various sensors and control devices.

6. Market-based accounting for the use of services as well as learning when others consume services exposed by the given provider. This may also benefit from auctions for services where a service can be sold at an auction.

6.1.2 Domain Model of the System

Following and extending the requirements identified in the previous section, the author proposed a conceptual domain model for a middleware with main entities outlined as well as crucial relationships shown in Figure 6.1. Subsequently, definitions of the most important entities are presented. The model is then implemented within BeesyCluster as described in Section 6.1.3.

6.1.2.1 User

The model distinguishes *User*s who can register in the middleware using *Account*s. The latter can then be assigned to specific *SystemAccount*s on particular physical resources to which the user wants to gain access. The model allows *User*s to set up an account without registering system accounts, in which case such a user can only become a client of services published to them.

The *User* who has published a *Service* or *Service*s and makes the latter available to others becomes a *Provider*.

Furthermore, users can create and join *User Group*s which are defined as collections of *User*s. The users within a *UserGroup* can share *UserTask*s, exchange *Message*s, and draw and share *Design*s via *SharedSpace*s. It is also assumed that users can interact online via *ChatRoom*s (Figure 6.1).

6.1.2.2 Resources

In the model, a *Resource* is as either a physical or virtual entity within a computer system. The following types of resources can be distinguished:

1. *A PhysicalResource* that represents a physical component in the system. In this case, for the purpose of this work, the following classes are important:

   - *Computing*—corresponding to a processing resource such as powerful CPUs and possibly GPUs
   - *Storage*—storage systems which can be used by *Computing* resources to fetch data from and send output to
   - *Interconnects*—for modeling communication among components

2. *A VirtualResource* that is mainly *File*s stored on *Storage* resources such as disks and disk arrays.

The aforementioned resources are contained within *Cluster/Server*s which consist of *Computing*, *Storage*, and *Interconnect* resources. In the case of a server, it can be a CPU and a disk. In case of powerful HPC clusters, this can include thousands of CPUs, GPUs, disk arrays, interconnects such as InfiniBand, etc.

It should be noted that *Cluster/Server*s use virtual resources in several ways:

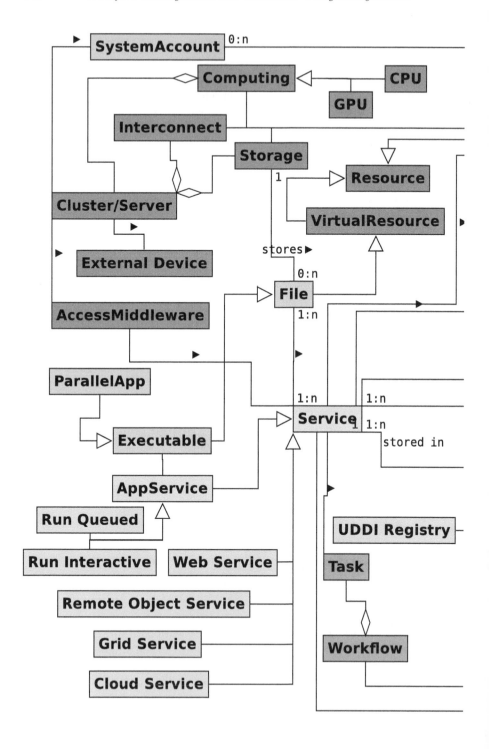

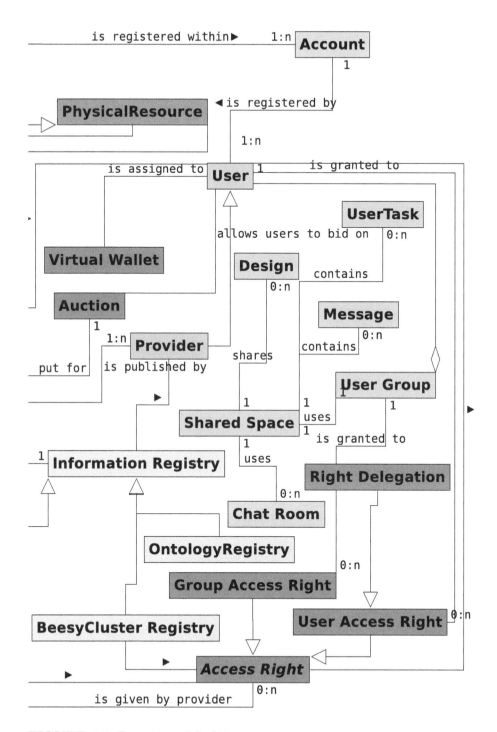

FIGURE 6.1: Domain model of the system.

1. Regular files stored within *Cluster/Servers* can contain data, either input or output for services.

2. Some files can be used as resources within specific standards of services. For instance, the WSRF specification [53] allows a service to operate on a resource, which can be a file, and store there data corresponding to the current state.

3. Executables stored within *Cluster/Servers* can be run as either sequential or parallel (e.g., using MPI) applications.

Each *Cluster/Server* holds several *SystemAccounts* that correspond to user accounts in the operating system installed on the given cluster or server. Each *User* has a single account in the middleware which is associated with zero or more *SystemAccounts*. In the former case, the user may be a client of services published by others.

It is assumed that the *VirtualResource* is *managed* by its owner since the resource has been installed or is located on their account. Thus the owner is responsible for correct installation, existence of libraries required to run the *Executable*, or sufficient operating system privileges to view or edit the file.

An *Executable* may be a:

- sequential process; in fact, console-based applications reading from standard input and writing to standard output/error streams will be most useful for integration and can be integrated as parts of complex workflows.

- parallel application within specific MPPs using MPI, PVM [218] or other libraries or environments.

- parallel application spanning clusters using technologies like MPICH-G2 [126], PACX-MPI [128], BC-MPI [59] or others.

Additionally, *Executables* may be intentionally designed to control *ExternalDevices*. This is especially useful for remote control of devices that can be attached to computers, reading data from sensors installed on, e.g., smartphones accessible from an *Executable* using Bluetooth, etc. Such applications can be exposed as services.

6.1.2.3   Services

Each *User* may expose *Services* that will be made available to other users for use. There are several types of *Services* that can be defined:

1. Executables that can be either sequential or parallel applications; in the latter case queuing systems such as PBS, LSF, or LoadLeveler can be represented within an *AccessMiddleware* described next; it can be noted that requests handled by a human can be implemented as, e.g.,

an application that sends an email to a human and waits for a response that will be passed as output from the service in the system,

2. Standard Web Services [185], remote object service, grid service, cloud services, etc. as discussed in Chapter 2.

In this context, applications are services managed by their owners in a market of services.

AppServices are those that are associated with virtual *Resources*, i.e., *Files*. There can be two types of such services:

1. An *Executable* is installed on a cluster or a server by the provider.

2. An *Executable* is passed to a cluster or a server which selects the best physical resource, e.g., a node or nodes on which to run it. This is especially useful if a cluster or a server is actually deployed on a grid.

Services are run on behalf of the owner of the *SystemAccount* on which the virtual *Resource* is installed. This, in general, allows us to run a service on:

- the provider's account but in a sandbox, which prevents the client from gaining access to a shell and allows uploading of input data and fetching of output data only;

- a dedicated account set up by the *Cluster/Server*'s administrator.

This model is different from virtual accounts as used in grid systems, e.g., VUS in the CLUSTERIX project [119] but allows users to manage resources from their own accounts. Additionally, external *Services* are considered that do not use managed resources installed on a cluster or a server as indicated above but are generally available services offered by third parties from other locations not registered in the middleware as the clusters or servers. This refers to, e.g., Web Services, Web Services/WSRF or cloud services as proposed in Chapter 3.

AccessMiddleware represents a middleware or an additional software layer that is used to launch and control a service.

6.1.2.4   Access Rights

Providers assign *Access Right*s to published services, to both individual *User*s and *UserGroup*s to which they wish to grant access. Access can be granted for a limited time within a day, for selected work days, and within a specified time limit.

6.1.2.5   Virtual Payments

It must be noted that clients are charged for the use of paid services published by providers. An amount equal to the price is then deducted from their *Virtual Wallet*s. Additionally, services may be put on auction where the winning bidder is charged and is then able to use the given service.

6.1.2.6   Information Registry

It is assumed that providers who have published services provide relevant descriptions which are placed in an information registry. Such a registry can be searched for services meeting given criteria. The following types of registries are considered:

1. a UDDI registry for exposing services to clients who do not have accounts in the middleware,

2. an embedded database in the system for high-performance access compared to SOAP-based access to UDDI,

3. an *OntologyRegistry* that contains an ontology in the domain of the services (e.g., a particular field of science or business). It can be used to couple requests for services with service descriptions through semantic connections in the ontology. Such a solution is discussed in Section 2.1.2.

6.1.2.7   Workflows

Finally, the model allows the user to aggregate services available to them from various providers using various usage policies into complex *Workflows* as well as execute them. A *Workflow* is composed of *Tasks* to which *Services* are mapped. It should be noted that providers can revoke the right to run the given service at any time.

6.1.3   BeesyCluster as a Modern Middleware for Distributed Services

The model proposed in the preceding sections has been implemented within the BeesyCluster system presented in this chapter. The architecture, main features, as well as modules are introduced. BeesyCluster is a middleware that allows registered users to access system accounts on various resources such as servers and clusters. The access is possible through either WWW (HTTPS) or Web Services (SOAP/HTTP(S)) [69]. The user is able to manage files on system accounts they have access to, and edit, compile, and run either sequential or parallel applications. BeesyCluster itself maintains several open SSH sessions with clusters and servers to run applications and manage files.

Currently, BeesyCluster [4] is deployed at the Faculty of Electronics, Telecommunications and Informatics, Gdansk University of Technology, Poland for management of the university clusters, servers, and laboratories including the resources of the Academic Computing Center.[1]

---

[1] http://task.gda.pl/centre-en/

6.1.4 Functions and Features

BeesyCluster provides an environment with the following:

- *Single sign-on and access to many distributed servers, clusters, and the services installed on them*; by logging into the BeesyCluster middleware, the user gains access to:

    1. File systems on the accounts registered by the user on the servers or clusters previously approved by BeesyCluster's administrator. Each user may submit a request for inclusion of new servers or clusters by providing their names and corresponding IPs. The administrator can then approve or reject the request. The system accounts can be registered within each BeesyCluster user account by selection of a server or cluster and provision of login/password pairs that are stored in an encrypted form by BeesyCluster. This registration scheme is a much simpler and faster process compared to often complex and time-consuming deployment of new computing nodes in grid systems [94], [209]. Users just interested in acquiring access to ready-to-use services published by others may set up a BeesyCluster account just for that without obtaining access to owned accounts.

    2. Services either published by the user or made available by others to this user or one of the groups to which the user belongs.

- *Loosely coupled services managed by providers*—BeesyCluster users can easily do the following:

    1. Publish services based on existing applications in a few clicks or method invocations. Such an application becomes a BeesyCluster service visible and accessible within the system and outside Beesy-Cluster as a Web Service.

    2. Define and change QoS parameters as well as grant or remove privileges to either BeesyCluster users or groups at any time. Services can be defined within seconds based on the applications the user has at their disposal, either already available or custom developed. Access can be granted within specific periods to the user or the group.

    3. Run services in one of the following ways:
        - interactively, even using graphical interfaces,
        - using one of queuing systems registered in BeesyCluster such as PBS, LSF, LoadLeveler, etc.

       It is possible to run services published by others in one of two ways:
        - in a safe sandbox on the provider's account allowing the client to check status and view as well as copy results,

  – by a service allowing downloading of a binary executable or
    installation files (if allowed by the provider) and running the
    application on one of the accounts available to the consumer; in
    the latter phase the architectures of both systems must match.

- *Service registry*—BeesyCluster provides two types of registries that store
  information about services and providers:

  1. an embedded database for fast access to data within the system,
     especially when verification is needed if the client or clients have
     sufficient privileges to run multiple services;

  2. a UDDI registry using jUDDI [99] that stores data exported from
     the embedded database for access from outside of the system. Se-
     lected services are then available through the Web Service interface.

- *Accounting*—users can use their virtual wallets for spending on services
  published by others. Likewise, users can earn when their own services
  are consumed by others.

6.1.5   Architecture and Core Modules

Conceptually, BeesyCluster consists of the middleware as well as interfaces
which allow access to the functions offered by the middleware (Figure 6.2).

Users usually focus on the functions available via the WWW interfaces
and BeesyCluster can be seen as an access portal to a network of resources
(clusters/supercomputers/PCs) from this perspective. Similarly, any geographically distant program can invoke BeesyCluster's functions via Web Services in which case the logic is performed by the middleware.

From the architectural point of view, BeesyCluster's middleware corresponds to technologies like Globus Toolkit, while the presentation layer corresponds to functions provided by high-level interfaces like Migrating Desktop
[136] in CrossGrid and a visual interface [220] in CLUSTERIX.

Since BeesyCluster has been implemented using the Java EE technology,
which itself imposed the multitiered architecture, particular modules of Beesy-
Cluster can be placed in appropriate layers as shown in Figure 6.2. The most
important modules include:

- *Presentation layer:*

  – *Cluster Commander (CC)*—for servlets either allowing user input
    or display of results to the user.
  – *Teamwork Support (TS)*—allowing definition of groups and work-
    ing within groups of users including definition of project time-
    lines, milestones, exchanging messages, sharing files, using chat and
    shared whiteboards; servers running out of scope of the Java EE
    server are used.

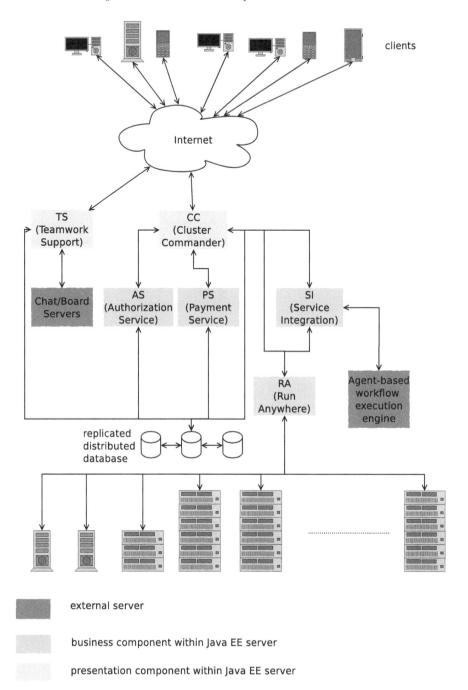

FIGURE 6.2: BeesyCluster's architecture with key components.

- *Middleware—business layer:*

  - *Authorization Service (AS)*—for verification of access of users and groups to resources, storing information about users, resources, privileges, delegations, etc.

  - *Payment Service (PS)*—handling access to paid resources with accounting.

  - *Run Anywhere (RA)*—a proxy for accessing and launching commands on distributed servers and clusters connected to the system using SSH; it is also the low-level layer for transfer of files between the BeesyCluster server and external resources.

  - *Service Integration (SI)*—for definition of workflow applications that can be built out of BeesyCluster services and subsequently executed; this module is discussed in Section 6.2 along with the internal and external execution engines.

- *Data/EIS layer*—BeesyCluster uses a MySQL database through the JDBC interface.

Figure 6.3 presents the actual deployment of BeesyCluster at Gdansk University of Technology and the servers, clusters, and laboratories registered in the system. It should be noted that in case of many clients, several Beesy-Cluster servers can be set up for parallel handling of multiple client requests. Additionally, the database itself can be configured to use multiple servers with replication [193].

6.1.6   Security

The authorization mechanism in BeesyCluster stems from the assumption that the provider who registers a new computing node in the network is not required to install any additional software on the system accounts on the computing node from which services may be published. This implies that a common and secure means of accessing such an account must be used. This must be present in a variety of systems, which is possible in the case of an SSH server allowing secure sessions. This, in turn, implies that users who wish to use BeesyCluster trust the middleware by disclosing their login credentials (login/password) to BeesyCluster, which stores these in an encrypted form in the database. Such credentials are used only by the BeesyCluster logic to access accounts and run services on behalf of their owners. Thus, BeesyCluster computing resource communication uses SSH as shown in Figure 6.2. The client-BeesyCluster communication uses HTTPS, which can be used both for WWW (Section 6.1.7.1) and Web Service invocations (Section 6.1.7.2).

The client is authenticated in BeesyCluster by the login/password pair when they first log in to their BeesyCluster account. This is true both for

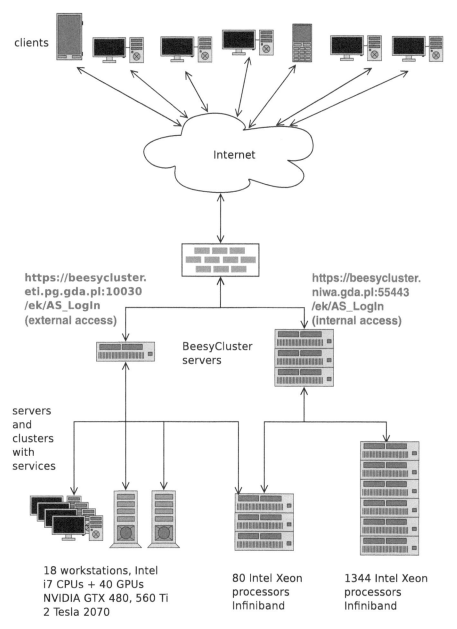

FIGURE 6.3: Deployment of BeesyCluster at Academic Computer Center and Faculty of Electronics, Telecommunications and Informatics, Gdansk University of Technology, Poland.

HTTPS communication (WWW) as well as Web Services. In both cases, after having logged in, the user receives an authenticator which is an encrypted token containing the user identification number and expiration date and possibly restrictions on the set of services the user is allowed to perform using this ticket. Thus, only BeesyCluster is able to decrypt the ticket and validate this information. Unauthorized change of the authenticator is not possible as will be instantly discovered during decryption. This method is typical of several other systems or standards, in particular UDDI, in which a similar mechanism with a token is used to allow publishers add new or update information in the information registry.

In the case of WWW information exchange, the BeesyCluster server writes the authenticator as a cookie on the client machine and subsequently fetches and validates the authenticator when the user visits following web pages.

In the case of Web Service communication, there is a dedicated method for logging in which takes the login and the password as arguments and returns the authenticator, which must be passed to all subsequent Web Service calls, in particular calling various services. BeesyCluster is also able to validate the user.

6.1.7   Running Services in BeesyCluster

In BeesyCluster, services assume input parameters, process input data, and produce output results. These can include:

- tasks run interactively (with user handling input/output while the task is running),

- parallel tasks queued on clusters using LSF, PBS, or launched with `mpirun`,

- editing/viewing files,

- any services added to the system that can operate on any of the resources attached to the clusters/PCs attached to the system. A student/industry laboratory can feature network cameras serving pictures taken by an embedded WWW server. The administrator of such a device (having login/password) can write a collection of simple C/Java programs which allow viewing of the images, rotate the lens, zoom, etc., and define proper privileges for users to invoke them. These are then made available as services in BeesyCluster.

A user, having logged in via the WWW interface, can run either their own services or services made available by others to which the user has been granted privileges (no further authorization is required after logging in). The user can access services via a list of available services, upload input files, run, browses download, and copy results between clusters/PCs to which they have access.

The BeesyCluster middleware verifies whether the user has access to the services they try to invoke. In particular, for each user BeesyCluster controls the spaces for uploaded input data for called services as well as dedicated directories where output data for particular invocations of services are stored. The user has access via WWW (Section 6.1.7.1) or Web Services (Section 6.1.7.2, [69]) to these spaces.

A service running outside of BeesyCluster can launch a service via Web Services in which case authorization analogous to WWW logging is required [69].

In BeesyCluster, it is also possible to install and run an application on one of the dedicated user accounts available on clusters or servers attached to the system [124]. This effectively implements searching for suitable and optimal (e.g., in terms of performance or performance/power consumption) resources much like in traditional grid systems where hardware is found for the user application.

6.1.7.1 WWW Interface

The presentation layer of BeesyCluster, incorporating Java servlets, offers users all functions described in Section 6.1.4. Specifically, management of files, running commands and publishing applications as services to be run subsequently are outlined in a use case diagram in Figure 6.4.

Following logging into BeesyCluster using the login and password, the following sequence presents steps necessary to develop, compile and queue a parallel MPI task with custom input files on a system account available to the user:

1. BeesyCluster's file manager allows the user to browse the files sytems on system accounts on clusters available to them. The Web browser, by remembering past inputs in the command text box, effectively implements the history of functions.

2. The system allows running an application interactively within a web browser, in which case the user has access to a command-line console. Alternatively, queuing can be launched from a context menu. It is worth noting that the user does not interact with the queuing system, in fact does not know what queuing system is used on the given cluster. The user only specifies the input file(s) in a text box which must be uploaded to the system, the required number of processors, and notification settings. BeesyCluster [100] selects the best queue on the cluster that meets these criteria unless the user specifies the queue as well.

3. The user can then view statuses of previously submitted tasks.

4. Finally, the user can fetch results of the task which are generated to a dedicated directory on the system account.

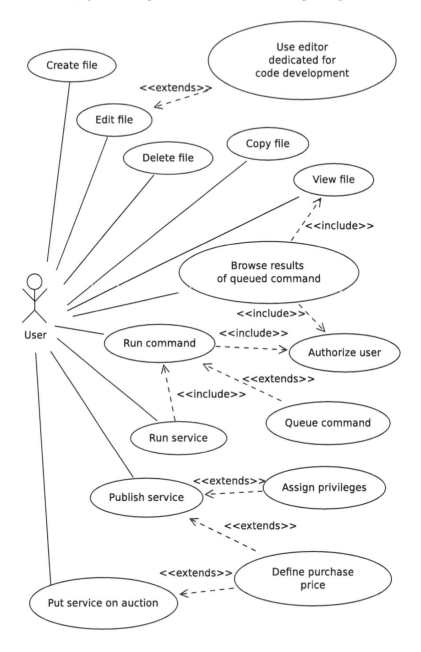

FIGURE 6.4: Use case diagram for BeesyCluster's WWW interface.

6.1.7.2   Web Service Interface

Similar to the steps outlined for the WWW interface, the same functionality of running a command on the given system account can be accessed using Web Services. In this case, running a task without the queuing system is presented for simplicity. Listing 6.1 presents a template code which should be used to run a command on a system account on a cluster:

Listing 6.1: Client code to invoke a command via BeesyCluster.

```
import pl.gda.pg.eti.beesycluster2.*;
import as.interfaces.auth.*; import as.interfaces.sign.*;

4 // obtain a stub to an authorization service in BeesyCluster
  // that will allow logging in and getting an authenticator

  ASService service = new ASServiceLocator();
  AS asstub = service.getAS();
9
  try { // obtain a login agent and signer module ids

        LoginAgentDescription[] lad=asstub.listLoginAgents();
        SignerDescription[] sd=asstub.listSigners();
14
        // log into BeesyCluster with the same login/password
        // used for the WWW interface

        String[] authenticator = asstub.logIn(new String[] {"<
            login>","<password>"},
19      lad[0].getID(),sd[0].getID());

        // obtain a stub for the service that allows
        // calling commands on a given cluster
        // accessible to the BeesyCluster server
24
        KCWSService kcwsstub=new KCWSServiceLocator();
        KCWS asstub1=kcwsstub.getKCWS();

        // run <command> on the cluster with a given id
29      // and display result
        System.out.println(asstub1.runCommand(auth,3,<command>))
            ;


  } catch (IOException e) {
    // handle exceptions
34    ...
  }
```

1. getting an id of a login agent object;

2. getting an id of a signer object;

3. logging the user in: the input parameters include the login, password, the ids of the login and signer objects; an authenticator is returned which needs to be passed to subsequent calls such as invocations of services in BeesyCluster or applications on accounts to which the user has access;

4. calling the actual command on a cluster or server registered in the Beesy-Cluster middleware; the parameters include the authenticator, the cluster and the command.

6.1.8   Publishing Services

BeesyCluster is mainly about a distributed environment in which users can publish and control access rights to their own resources. To achieve that, and also to be able to integrate services into workflows, users-owners must first make proper actions on resources available as services. The following steps are required:

1. The user who wishes to publish a service simply clicks on the share icon in the file manager.

   Subsequently, a unique name for the service as well as its parameters must be defined. The latter include: a link to a manual, whether the application should be queued, the minimum and maximum allowed number of processors for running the application, and whether these parameters can be changed by the user who will be granted access.

2. Subsequently, the cost of running the service needs to be defined (the service can be annotated as free of charge with the cost equal to zero).

3. Having confirmed publishing, the service the provider obtains confirmation of the result. Next, the owner must grant proper access rights to user(s) to whom the service is to be made available.

4. The client can now upload input files (formats presumably described in the manual downloaded in the previous step) as well as queue the service.

5. After the task has been queued and executed, the client can browse its results and copy them to one of their system accounts.

6.1.9   Performance of Web Service and WWW Interfaces

The first obtained performance result related to the Web Service interface (SOAP/ HTTPS) is the time needed to log in to BeesyCluster and subsequently invoke the `pwd` command:

- LAN network—both the client and the BeesyCluster server were in the same LAN, 85 ms,

- Broadband—the client accesses the server through a broadband connection to the Internet, 315 ms.

Regarding the hardware equipment a 2-GHz, Pentium4, with 1-GB RAM machine was used for the server while a 2.8-GHz, Pentium4 1-GB RAM laptop, both running Fedora ran the client code.

Subsequently, the times needed for various common operations using the WWW interface (HTTPS) are shown in Figure 6.5. Each operation was executed in two configurations, LAN and broadband, as defined above.

6.2   Workflow Management System in BeesyCluster

Monitoring of services, building models of complex tasks out of simple tasks with functional and non-functional specifications, assignment of services, scheduling, and execution and management of results, are typical tasks available in workflow management systems. This section presents functions, architecture, and implementation including novel ways of workflow execution built on top of the BeesyCluster system.

6.2.1   Workflow Editor and Execution Engine in BeesyCluster

6.2.1.1   Functional Requirements and Exception Handling

So far, the functions of the BeesyCluster middleware presented in Section 6.1.4 referred mainly to publishing owned services from user-controlled computing nodes with privileges set on a user or group basis by the service owner. Both the WWW and Web Service interfaces allow the client to upload input data, run the service, and fetch input results. The latter could then be defined as input for other services thus creating a workflow. This process has the following drawbacks, though, since it requires:

1. manual intervention from the client for input/output data management,

2. monitoring and awareness of the speeds of the clusters on which the

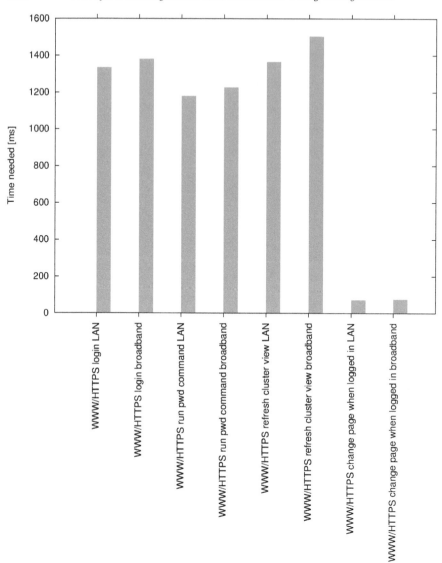

FIGURE 6.5: Performance of the WWW interface (HTTPS).

services run, as well as their costs for determining which services should be chosen for execution of the given task assuming certain cost/workflow time constraints are met.

For these reasons, the author designed and implemented a BeesyCluster module [66] which provides:

1. An intuitive interface for workflow creation by drawing a graph of tasks represented by nodes and task dependencies by edges.

2. Presenting users with a list of services available to them, possibly running on provider-owned computing nodes and published by various providers.

3. Easy assignment of services capable of executing workflow tasks.

4. Automatic generation of constraints according to the algorithms presented in Sections 4.2.1 and 4.4 allowing the user to edit and change constraints; the latter are generated taking the following into consideration

   • Input files specified by the user—the user specifies files on each of their system accounts registered in BeesyCluster; the module obtains the size of data automatically by querying the clusters.

   • Service costs obtained automatically from the BeesyCluster database—costs defined at the publishing time by the service owner; the cost can be changed at runtime.

   • Cluster performance.

5. Execution of workflows using any of the algorithms presented in Section 4.2 which optimizes data flowing to particular services, effectively choosing services for particular tasks; the execution module copies data automatically between services (possibly running on various clusters on various system accounts and published by various users) and runs services immediately or queues using a local queuing system if necessary.

These functional requirements are presented in a use case diagram in Figure 6.6.

Furthermore, the following exceptions are considered and solutions provided:

1. Workflow editing: If the client-BeesyCluster connection fails due to network problems or server crash, the client can reconnect to any of the other BeesyCluster servers to continue the process of workflow editing.

2. Workflow execution: The workflow mapping may be reevaluated if:

   • a node with a service to execute within the workflow becomes unavailable during workflow execution but before the service was started, a new mapping with an alternative service for the task can be chosen;

   • if the node becomes overloaded by other users even during the execution of the service, the service can be checkpointed using one of the methods described in [72] and a new workflow mapping can be

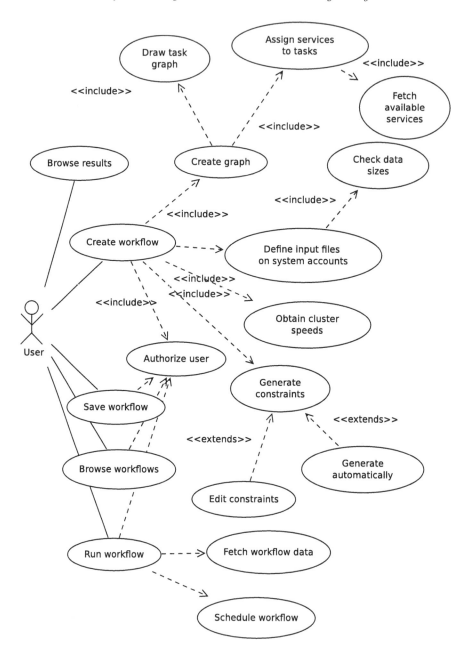

FIGURE 6.6: Use case diagram for BeesyCluster's workflow management system.

obtained; in this case it is assumed that the service either continues to run on the overloaded node or is checkpointed and migrated to another processor(s) of the node/cluster/system account.

6.2.1.2   Editing Workflows and Execution Engine

This section presents crucial classes and interaction schemes both for editing and execution of workflows composed of basic services made available by providers through BeesyCluster.

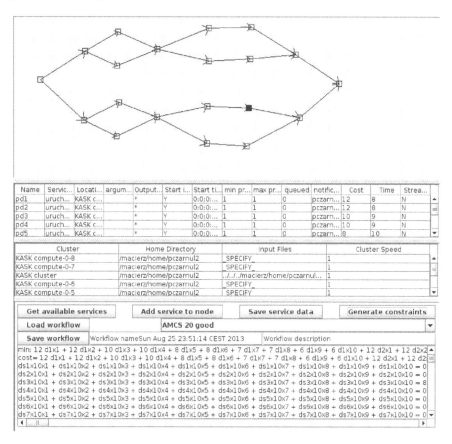

FIGURE 6.7: Workflow application editor in BeesyCluster.

Editing a workflow is realized through an easy-to-use editor implemented as a Java applet (Figure 6.7). Definition of a workflow requires the following steps, some of which trigger querying the BeesyCluster server for additional information:

1. The user refers to the `SIEditorServlet` which performs proper initialization, downloads an editor applet to the client browser, and starts it.

2. The applet connects to the BeesyCluster server using secure sockets and a new server thread is started for communication with this particular client.

3. Services available to the client (possibly from various providers) are downloaded and displayed.

4. Information regarding system accounts available to the user is downloaded.

5. The user draws a graph of tasks with intertask dependencies indicating which task(s) should be executed before others. This is in line with the formulation shown in Section 3.4.2. Before selecting a task, at least one service implementing the task must be selected.

6. The user can assign more services capable of executing a particular task from the list of available services.

7. After editing the graph has been finalized, the user needs to specify input files which will be used for the workflow. Any number of expressions meaningful to the shell can be defined for each system account registered for the user, and wildcards can also be used.

8. The user can update the speeds of the clusters as well as the costs of the services used when selecting the sizes of data which should flow to particular services (effectively selecting a service) even though these are automatically downloaded and set.

9. Before saving the workflow on the BeesyCluster server for further execution, constraints for the workflow for the algorithm from Section 4.2.1 are generated automatically. The constraints use the sizes of the input data specified by the user, the dependencies between tasks, assignment of services to tasks, clusters speeds, locations of services in terms of larger communication costs between services located on different nodes, and potential incompatibility of subsequent services published by different providers and thus added translation costs. The user is able to further modify the constraints.

10. Definition of the workflow is uploaded to the BeesyCluster server.

Furthermore, several classes encapsulating data concerning the workflow must be passed between the client and the server. These are shown in Figure 6.8 and include:

- *SIWorkflowData*—data concerning the workflow including the name, description along with constraints related to the workflow application including time, cost, etc.

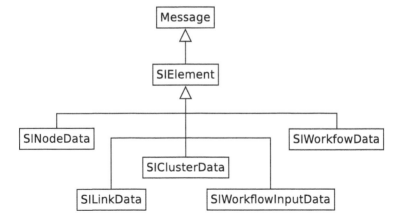

FIGURE 6.8: Hierarchy of classes for workflow management.

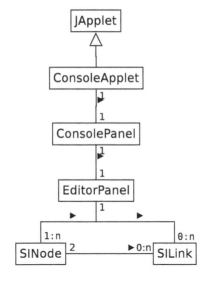

FIGURE 6.9: Classes for the applet running on the client side.

- *SINodeData*—data concerning the particular service; each service is described as a separate SINodeData with the same node identifier but with a different service index.

- *SILinkData*—links connecting nodes of the given source and target node identifiers.

- *SIWorkflowInputData*—concerns input data, i.e., input files specified by the user for a particular system account on a cluster.

- *SIClusterData*—used for querying the server about the cluster speed and the sizes of the input data specified by the user.

Key classes of the applet are shown in Figure 6.9. On the server side there is a **Server** thread querying clusters for the their speeds and sizes of specified input files through the **RA** module. Important to note is the **EditorPanel** for drawing graphs which builds the graph at runtime by creating **SINode** and **SILink** objects. Each **SINode** object contains references to all services (**SINodeData** objects) selected by the user, initially fetched from the server and possibly changed (arguments, desired numbers of processors, etc.) by the user. Final values are then returned and uploaded to the **Server** which stores the workflow data in the database.

Figure 6.10 presents the main components involved in editing of the workflow, including the applet, the **Server** with which the applet communicates over secure sockets. The **Server** uses **Resource**, **PayableRes** (resources that require payments), general methods of the **PS** and **AS** beans for fetching a list of available services, the **RA** module for querying the cluster for input data sizes and cluster speeds, and the following **SI*** beans for saving the workflow in the system database: **SIWorkflow** (stores data from **SIWorkflowData**), **SIWorkflowNode** (stores data from **SINodeData**, i.e., about services assigned to a particular workflow node), **SIWorkflowLink** (stores data from **SILinkData**), and **SIWorkflowInput** (stores data from **SIWorkflowInputData**). These beans store data in the database and provide object representations of the data stored in a relational database.

Workflow execution starts after the user has browsed available workflows created in the workflow editor and selected one for execution (Figure 6.11). This is done by the **ServiceIntegrationServlet**, which also initializes the execution of the workflow:

1. A new workflow instance **SIWorkflowInstance** is created.

2. A new workflow mapping solver (**WorkflowMappingSolver**) is created for the workflow selected by the user, workflow constraints are fetched, and the problem is solved using one of the algorithms described in Section 4.2. For the ILP algorithm, **lp_solve** [9] is used. In the model considered in Section 3.4.3 for which task data sizes are known in advance, the results include selected services and schedules. In the model considered in Section 4.4, resulting values include data sizes which should be sent to particular services in the workflow, effectively selecting the services. If the input data size is equal to 0, a corresponding service will not be activated.

3. All nodes of the workflow are fetched. For each node a new dedicated directory is created (through the **RA** module) on the system account on which the service was installed (`<home_dir>/<workflow_instance_id>-<node_id>`).

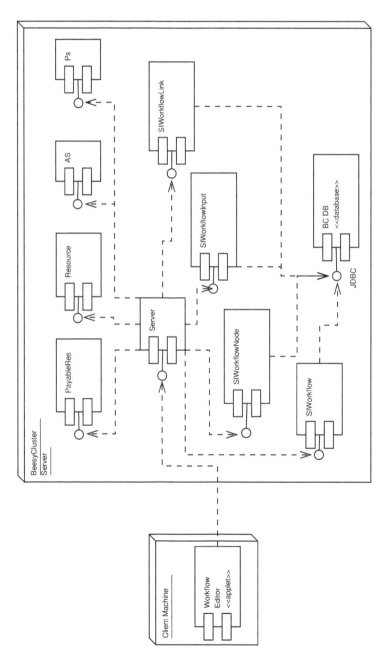

FIGURE 6.10: Deployment diagram with components for editing workflow applications.

Workflow 6 nodes AMCS	Workflow description	Run (optimize before run)	Run (dynamic optimization)	Run (full dynamic optimization)	Run (HEUR)	Run (GA)	Run (GAIN - budget constrained)	Run (layered)	View	Delete
AMCS 6 nodes 5 services each	Workflow description	Run (optimize before run)	Run (dynamic optimization)	Run (full dynamic optimization)	Run (HEUR)	Run (GA)	Run (GAIN - budget constrained)	Run (layered)	View	Delete
paper2010 6 nodes 5 service each	Workflow description	Run (optimize before run)	Run (dynamic optimization)	Run (full dynamic optimization)	Run (HEUR)	Run (GA)	Run (GAIN - budget constrained)	Run (layered)	View	Delete
New workflow	Workflow description	Run (optimize before run)	Run (dynamic optimization)	Run (full dynamic optimization)	Run (HEUR)	Run (GA)	Run (GAIN - budget constrained)	Run (layered)	View	Delete
3 nodes - test streaming	Workflow description	Run (optimize before run)	Run (dynamic optimization)	Run (full dynamic optimization)	Run (HEUR)	Run (GA)	Run (GAIN - budget constrained)	Run (layered)	View	Delete
3 nodes 1	Workflow description	Run (optimize before run)	Run (dynamic optimization)	Run (full dynamic optimization)	Run (HEUR)	Run (GA)	Run (GAIN - budget constrained)	Run (layered)	View	Delete
streaming 4 paths length 2	Workflow description	Run (optimize before run)	Run (dynamic optimization)	Run (full dynamic optimization)	Run (HEUR)	Run (GA)	Run (GAIN - budget constrained)	Run (layered)	View	Delete
rozdzielenie 179	Workflow description	Run (optimize before run)	Run (dynamic optimization)	Run (full dynamic optimization)	Run (HEUR)	Run (GA)	Run (GAIN - budget constrained)	Run (layered)	View	Delete
capacity test 80 files - workflow 180	Workflow description	Run (optimize before run)	Run (dynamic optimization)	Run (full dynamic optimization)	Run (HEUR)	Run (GA)	Run (GAIN - budget constrained)	Run (layered)	View	Delete
Workflow 181 - capacity constraints 4 paths 2 node	Workflow description	Run (optimize before run)	Run (dynamic optimization)	Run (full dynamic optimization)	Run (HEUR)	Run (GA)	Run (GAIN - budget constrained)	Run (layered)	View	Delete
2 paths - image processing - no streaming	Workflow description	Run (optimize before run)	Run (dynamic optimization)	Run (full dynamic optimization)	Run (HEUR)	Run (GA)	Run (GAIN - budget constrained)	Run (layered)	View	Delete
2 paths - improved no streaming	Workflow description	Run (optimize before run)	Run (dynamic optimization)	Run (full dynamic optimization)	Run (HEUR)	Run (GA)	Run (GAIN - budget constrained)	Run (layered)	View	Delete
AMCS good 40	Workflow description good 40	Run (optimize before run)	Run (dynamic optimization)	Run (full dynamic optimization)	Run (HEUR)	Run (GA)	Run (GAIN - budget constrained)	Run (layered)	View	Delete
AMCS good 102	Workflow description	Run (optimize before run)	Run (dynamic optimization)	Run (full dynamic optimization)	Run (HEUR)	Run (GA)	Run (GAIN - budget constrained)	Run (layered)	View	Delete

FIGURE 6.11: Servlet with previously defined workflow applications.

Launched workflows

Instance number	Name	Status	Execution time	Cost	a*Exec Time+sum b*Cost	alg exec time	View	Delete
7800	AMCS 20 good	FINISHED	1387.0 s	cost=1004.0	2391.0	algTime=20794	Visualize	Delete
7801	AMCS 20 good	FINISHED	1345.0 s	cost=980.0	2325.0	algTime=10982	Visualize	Delete
7802	AMCS 20 good	FINISHED	1375.0 s	cost=960.0	2335.0	algTime=21069	Visualize	Delete
7803	AMCS 20 good	FINISHED	1346.0 s	cost=968.0	2314.0	algTime=21071	Visualize	Delete
7804	AMCS 20 good	FINISHED	1305.0 s	cost=1052.0	2357.0	algTime=21118	Visualize	Delete
7805	AMCS 20 good	FINISHED	1354.0 s	cost=836.0	2190.0	algTime=8840	Visualize	Delete
7806	AMCS 20 good	FINISHED	1410.0 s	cost=816.0	2226.0	algTime=8856	Visualize	Delete
7807	AMCS 20 good	FINISHED	1381.0 s	cost=808.0	2189.0	algTime=9132	Visualize	Delete
7808	AMCS 20 good	FINISHED	1415.0 s	cost=792.0	2207.0	algTime=9068	Visualize	Delete
7809	AMCS 20 good	FINISHED	1488.0 s	cost=792.0	2280.0	algTime=9040	Visualize	Delete

FIGURE 6.12: Servlet with previously run workflow instances.

4. All initial nodes of the workflow are fetched (`SIWorkflowNode`).

5. All input files for all system accounts of the user are fetched.

6. For each initial node, certain input files (determined by the sizes returned from `WorkflowMappingSolver`) are transferred to the newly created input directories.

7. Execution of the workflow is initiated using JMS invoking the `onMessage` method in `SIMessageBeans`. The message sent includes the workflow instance identifier, the identifier of the service to be executed, and the identifier of the user.

Within the application server, there is a new `SIMessageBean` created for each service. Each node fetches its successors using the `SIWorkflowNode` interface. Since all executing nodes are obliged to create their `SIWorkflowNodeInstance`s after they are finished, each node can determine whether it has been the last predecessor of the given successor node. This is done within a static method in which the node marks that it has finished the execution of the service and checks whether there are any predecessors left. If not, the node sends a JMS message invoking a new `onMessage` method in a new bean corresponding to the successor.

The execution of each workflow node is performed in the following steps:

1. Data from the message is decoded so that the `onMessage` method identifies the workflow, node and user identifiers.

2. Access verification to the service is performed.

3. Files in the node directory are fetched as the input files for the given service.

4. For each file, depending on the service type (interactive, queued application, Web, WSRF Service etc.), the service is launched with the input file as an argument, possibly with additional arguments specified by the user in the workflow editor; matching input names and putting into arguments is possible.

5. For each successor and depending on the sizes of data which should be sent to following services, files are copied to node directories on the system accounts where following services were installed.

6. The bean notifies successors for which it is the last finishing predecessor.

Similar to Figure 6.10, Figure 6.13 presents the main components used for execution of the workflow, namely the `ServiceIntegrationServlet` initiating the execution, invoking `WorkflowMappingSolver` for workflow mapping, `SIMessageBean` for execution (using `RA`) of workflow services and passing data between services, `SI*` beans for fetching information on services to execute, and `SI*Instance*` beans for saving information on the workflow being executed.

It is possible to define when the particular service will be started. If not immediately, proper constraints on the startup time of the given node can be added to the constraints, i.e., $t_x^{st} \geq$ time for processing of task t_x not to begin before specified time. The required number of processors, and notification options can also be added.

The BeesyCluster workflow management system contains implementations of the algorithms presented in Section 4.2 that can be invoked [66]:

1. before the workflow application is started for the available services,

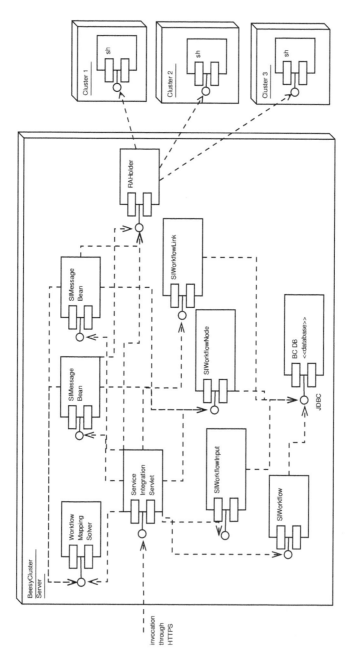

FIGURE 6.13: Deployment diagram with components for workflow execution.

2. during workflow execution if new services have become available or a previously chosen service has become unavailable.

Additionally, it is possible to introduce new algorithms and their implementations by just extending class `WorkflowMappingSolver`. The only other update necessary to make use of a new algorithm is to make the bean responsible for execution of a task (`SIMessageBean`) and a starting servlet (`SIIntegrationServlet`) aware of the new implementation and using the latter.

6.2.2   Workflow Execution

After particular services have been selected for a workflow application as outlined in Chapter 4, the workflow is executed by an execution engine. The execution engine performs the following tasks:

1. executes particular workflow tasks by launching services selected by the `WorkflowMappingSolver`, and

2. constantly monitors quality metrics of:

   - available services,
   - the availability of the existing services,
   - new potential service candidates for workflow tasks.

   The latter may trigger workflow rescheduling for the remaining part of the workflow.

Irrespective of the scheduling optimization, workflow execution may be performed using various architectures:

1. *Centralized*—one designated server holds data about services, launches particular services, and acts as an intermediary in passing output data from a service as input data of successive services. It should be noted that although processing is centralized, it can be parallel, e.g., if the server code runs on multiple multicore processors. The parallel engine implemented in BeesyCluster using message-driven EJBs can make use of more and more available cores of today's processors. This execution method uses the representation described in the previous section.

2. *Distributed*—coordination of execution of many workflow tasks, especially those executed in parallel, can be performed in parallel by distributed software agents in a distributed system. In this case, description of the workflow is exported to BPEL [121] and the agents can call services directly.

6.2.2.1   Multithreaded Centralized

The author tested the workflow execution engine using message-driven Enterprise Java Beans within a Java EE server to detect to what degree the engine is able to make use of available computing cores (Figure 6.14). The testbed workflow application shown in Figure 6.15 was used which processes a large set of input images through parallel paths of the workflow. Within each path, images are processed in a pipelined fashion. Sending output images from each workflow node to a following node is handled by the BeesyCluster server. This means that multiple communication actions between various pairs of workflow nodes overlap and can be in fact executed in parallel by the server. This is possible since each workflow task is handled by a separate message-driven EJB in the Java EE server. The beans can use the available processing cores in the server.

The parameters of this test case are as follows. The application processes 135 input images, each 12–15 MB in size. Each workflow path consists of three stages: conversion of a RAW image to TIFF, normalization, and resizing with quality adjustment. The details of this application are further described in Section 7.2.2. BeesyCluster ran on a server with Intel Xeon CPU 2.80-GHz processors with HyperThreading for a total of 8 logical processing cores and 4 GB of RAM. Results are shown in Figure 6.16, which clearly shows that the workflow execution time benefits from parallelization of the workflow execution layer. This is possible thanks to parallel execution of the components on the Java EE platform.

6.2.2.2   Distributed Execution Using Software Agents

This approach allows for further optimization of parallel execution of a workflow application through the following mechanisms:

1. Parallel execution of various workflow tasks and parallel handling of communication between pairs of workflow tasks by many agents working in parallel. This is equivalent to many EJBs working within a Java EE server in the solution presented in Section 6.2.2.1.

2. Optimal placement of agents handling execution of parts of a workflow in a distributed environment. This is an additional advantage compared to the centralized parallel approach since the agents can migrate to places from which access to particular services used by the workflow is much faster than from the fixed location of the Java EE server. Additionally, the agents can migrate to a new location to minimize transfer times of output/input data between services chosen to execute subsequent tasks.

In order to accomplish such a solution, a new parallel workflow execution engine, BeesyBees, using software agents for BeesyCluster was designed and implemented [78, 77]. After a workflow application is designed within the BeesyCluster workflow editor, it is saved on the server side in a database used

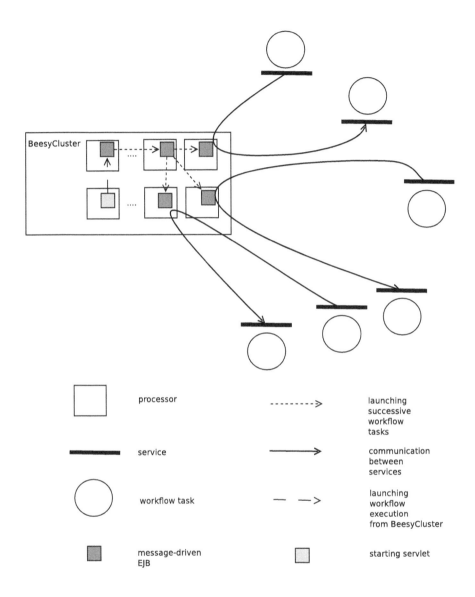

FIGURE 6.14: Multithreaded centralized execution in BeesyCluster.

by the Java EE server where BeesyCluster runs. Upon starting the execution using agents, the following actions take place [78]:

1. The BeesyCluster server invokes a dedicated Web Service that transfers the execution request to a GatewayAgent.

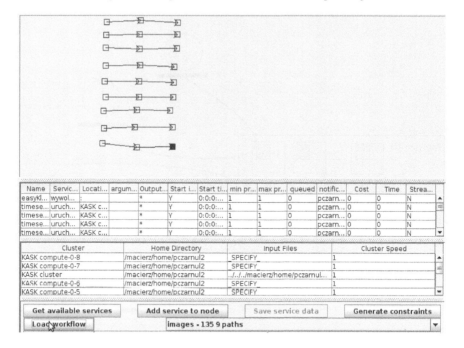

Name	Servic...	Locati...	argum...	Output...	Start i...	Start ti...	min pr...	max pr...	queued	notific...	Cost	Time	Strea...
easyKl...	wywol...	:	*	Y		0:0:0:...	1	1	0	pczarn...	0	0	N
timese...	uruch...	KASK c...	*	Y		0:0:0:...	1	1	0	pczarn...	0	0	N
timese...	uruch...	KASK c...	*	Y		0:0:0:...	1	1	0	pczarn...	0	0	N
timese...	uruch...	KASK c...	*	Y		0:0:0:...	1	1	0	pczarn...	0	0	N
timese...	uruch...	KASK c...	*	Y		0:0:0:...	1	1	0	pczarn...	0	0	N

Cluster	Home Directory	Input Files		Cluster Speed
KASK compute-0-8	/macierz/home/pczarnul2	SPECIFY	1	
KASK compute-0-7	/macierz/home/pczarnul2	SPECIFY	1	
KASK cluster	/macierz/home/pczarnul2	../../../macierz/home/pczarnul...	1	
KASK compute-0-6	/macierz/home/pczarnul2	SPECIFY	1	
KASK compute-0-5	/macierz/home/pczarnul2	SPECIFY	1	

Get available services	Add service to node	Save service data	Generate constraints
Load workflow	images - 135 9 paths		

FIGURE 6.15: Exemplary workflow application for testing parallel efficiency of the workflow execution engine.

2. A StatusAgent approves the workflow to be launched and controlled in the JADE environment or rejects it.

3. A StatusAgent launches TaskAgents that will be responsible for parallel execution of the workflow application.

4. TaskAgents negotiate which workflow tasks are handled and launched by which TaskAgents. Each TaskAgent prepares a list of tasks it wants to execute with a corresponding matching score for each task. Such information is sent to other TaskAgents as proposals. If another agent has a lower matching score for the task, it approves its execution, otherwise it refuses to execute it and can propose to execute it by itself.

5. The TaskAgent responsible for execution of a service for task t_i performs additional optimization that aims at minimization of the data transfer time and consequently the execution of the workflow. The following steps are applied:

   (a) The TaskAgent checks if the service $s_{i\ sel(i)}$ selected for execution of the task is available. The service should have been selected previously by the WorkflowMappingSolver in BeesyCluster. If the ser-

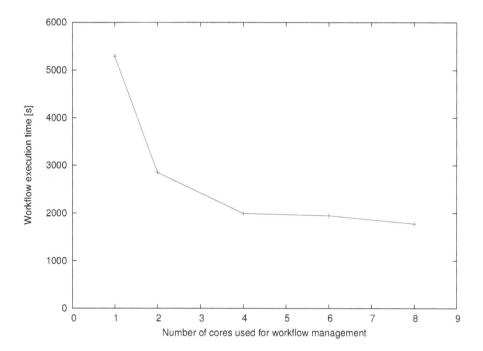

FIGURE 6.16: Impact of the number of cores used by the Java EE server on the workflow execution time.

vice is not available, the agent can find out and query the next best service given the optimization criterion as outlined in Section 3.4.2.

(b) The TaskAgent checks for which container c for which the transfer time of output data from the previous service $s_{p\ sel(p)}$ to service $s_{i\ sel(i)}$ will be the smallest since the container on which the agent runs will act as a proxy in communication. For this purpose, the following value should be minimized for the given preceding task t_p and container c:

$$t_{comm\ p\ c\ i} = t_{migration\ cnow\ c} + t_{startup\ p\ c} \qquad (6.1)$$
$$+ \frac{d_p^{out}}{B_{p\ c}} + t_{startup\ c\ i} + \frac{d_p^{out}}{B_{c\ i}}$$

where $t_{startup\ p\ c}$ and $t_{startup\ c\ i}$ correspond to startup times and $B_{p\ c}$ and $B_{c\ i}$ to bandwidths from the location of service $s_{p\ sel(p)}$ to container c and from container c to the location of service $s_{i\ sel(i)}$, respectively. $t_{migration\ cnow\ c}$ denotes the cost of migration of the agent from the current container $cnow$ to container c.

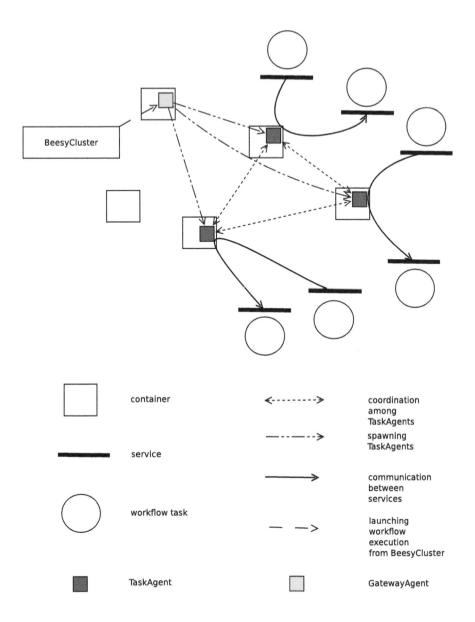

FIGURE 6.17: Parallel workflow execution using distributed software agents.

The author recommends extending this approach over all tasks t_p preceding task t_i and computing:

$$t_m = min_c\{max_p \, t_{comm \, p \, c \, i}\} \qquad (6.2)$$

and comparing with the value of $t_{comm\ p\ cnow\ i}$ for the current container for which the cost of migration equals 0. If $t_{comm\ p\ cnow\ i} < t_m$, then the agent stays where it is, otherwise it migrates to container c for which t_m has the minimum value.

In [78, 77], a workflow application with structure $(t \to t \to t)^3 \to t \to (t||t)$ was executed and tested. This workflow application can be treated as a template applicable to both scientific and business applications for which computations or business transactions with third parties can be executed in parallel, the results of which are integrated and products of the latter further distributed or processed in parallel.

It can be noted that separate agents are responsible for execution of the parallel paths of the workflow. Naturally, as in the centralized parallel case, the execution time is shortened by agents working in parallel. Additionally, data transfer times can be shortened if the agents are placed on containers closer to the services they are responsible for. In [78] three different scenarios were considered:

1. the worst possible placement of TaskAgents,

2. a random placement of TaskAgents,

3. the placement optimized using the aforementioned algorithm.

Figure 6.18 presents the average gain from an optimized placement of Task-Agents for the tested workflow applications compared to a random placement. At the same time, average difference of workflow execution times between the random and the worst possible placement is given as reference. It can be used to predict potential gain from distribution of workflow execution management and minimization of communication times by distributed agents.

Figure 6.19, on the other hand, shows gain that can be seen from increasing the number of containers, i.e., nodes, in the system to which TaskAgents can migrate. For a given number of TaskAgents, this allows agents to get closer to services. In a configuration with a single container, the container has the same distance in terms of communication costs to all nodes with services and invokes the latter. In the configurations with 4 and 8 containers, a TaskAgent on each container invokes services on 3 or 2 nodes with services, respectively, to which the distance from it is equal. In the 4- and 8-container configurations, the distance between a TaskAgent and each node with a service that it invokes is half the distance between the container and service nodes in the 1-container configuration.

It can be seen that for each following configuration and the number of containers to which TaskAgents can migrate, the total workflow execution time is smaller. This is because the agents can move to containers which are closer to the services between which the agents act as proxies for communication. Consequently, the execution time becomes smaller.

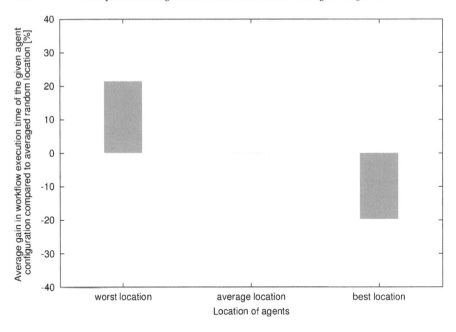

FIGURE 6.18: Gain from various TaskAgent location strategies in workflow execution.

Potentially a more optimal placement of agents compared to a centralized server in this solution can result in shorter communication and consequently workflow execution time. Additionally, failures of individual agents do not cause failure of the workflow application as in such a case another TaskAgent is launched that takes over execution of a service. Failure of a service is handled by the corresponding TaskAgent that can reselect a new service as in the centralized solution.

6.2.2.3 Distributed Using Multiple Application Servers

In some cases, the workflow scheduling algorithm may schedule parts of a workflow (e.g., a path or a larger subworkflow) in a specific geographical location, i.e., services assigned to particular tasks in the workflow part are in close geographical proximity to minimize communication. If this is the case, the following strategy is possible in workflow management in BeesyCluster to further maximize performance (Figure 6.20):

1. Split the original workflow application into partitions in which workflow tasks are additionally marked with partition numbers. This can be based on the communication proximity of services communicating in each subworkflow.

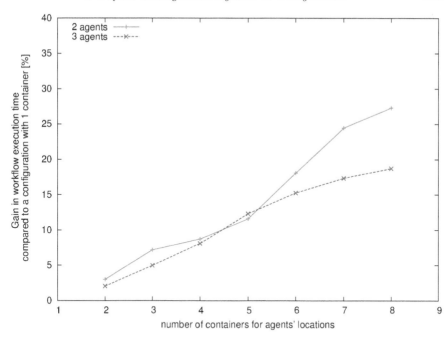

FIGURE 6.19: Gain from the number of containers for TaskAgents' locations in workflow execution.

2. Spawn a subworkflow or subworkflows as logically distinct workflow applications in BeesyCluster and launch on respective Java EE clusters in close proximity the service the workflows are using.

3. Gather results of the workflows on the original Java EE server from which subworkflows were spawned.

In this case, a subworkflow can be spawned on the particular Java EE using a remote EJB invocation. Alternatively, Web Service calls or even a standard GET invocation passed through HTTPS can be used for synchronization across partitions.

Note that it is possible that a following task needs to be activated after several other tasks have finished. Some of these may be run in a different Java EE cluster. In this case, a node in a different Java EE cluster can check in the database if it is the last predecessor of the following task. If this is the case, such a task initiates the following task by invoking an EJB remotely [168] (or by calling a Web Service) in the other Java EE cluster that initiates a new bean responsible for execution of a following task. The same method of synchronization is used when a subworkflow is synchronized with other subworkflows on a task that is supposed to be managed in another Java EE cluster.

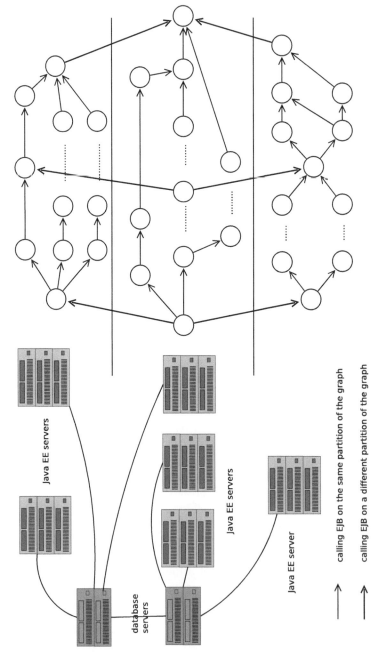

FIGURE 6.20: Parallel workflow execution using multiple application servers.

Note that the target Java EE server may have access to the location of input data to a particular workflow application as the database with access data can be shared among Java EE instances (Figure 6.20). All Java EE servers can use the same database system, which may be composed of several servers. This method ensures high, both intra- and interserver, scalability of workflow application execution.

The current implementation of the workflow management in BeesyCluster supports handling of failures of services [66] through successive launching of the service with an imposed timeout. If not succeeded, the service is assumed to be inaccessible or unable to return results and rescheduling is triggered using the `WorkflowMappingSolver`.

Chapter 7

Practical Workflow Applications

This chapter presents selected approaches and patterns in modeling practical workflow applications using the proposed models. Then practical workflow applications from various fields are presented along with expected and obtained performance results. It should be noted that the presented workflow applications can be of a specific nature such, as scientific—numerical computations (Section 7.2.2), business—production (Section 7.3) but also involving both business and scientific types of services, e.g., in a workflow for planning interdisciplinary projects (Section 7.4) or online translation on HPC resources (Section 7.5.3).

7.1 Exemplary Workflow Patterns and Use Cases

The model of a workflow application as well as algorithms for workflow optimization presented in Chapters 3–5 are general in nature and applicable to many application fields. It is the invention of the user and programmer how the proposed modeling constructs can be used to the benefit of a particular application. In this section, examples of general patterns using these modeling possibilities are presented. The following sections proceed with presentation of particular workflow applications that incorporate these patterns.

Each of the following patterns assumes processing data by services assigned to a task and optimization of a predefined goal with possibly other constraints as proposed in Section 2.2.5. The patterns may constitute a part of a real workflow application. In particular, the following patterns can be distinguished:

1. *Service selection pattern* (Figure 7.1)—this typical pattern assumes that each workflow task is assigned data of a predefined size. The goal is to select a service for each task such that a given goal is optimized.

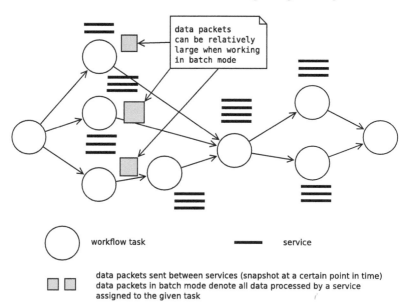

FIGURE 7.1: Service selection pattern.

2. *Service selection with data streaming tasks pattern* (Figure 7.2)—in this pattern, tasks are assigned data of predefined size. Some tasks can process data in streams. This means that we assume input data is given

as several data packets. If a given task processes data in the streaming mode and a data packet arrives at a service instance in this particular task, it is processed and output is forwarded immediately (presuming storage constraints allow this) to services assigned to following tasks. In the same workflow there may be some tasks that work in batch-processing mode if they require all previous outputs before proceeding. The goal is to select a service for each task such that a given goal is optimized.

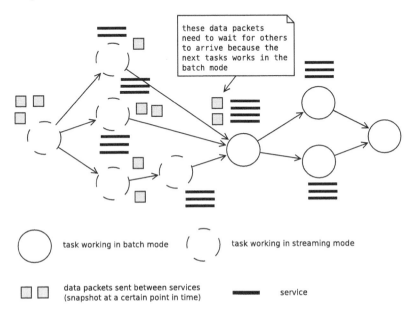

these data packets need to wait for others to arrive because the next tasks works in the batch mode

○ task working in batch mode   (⁻) task working in streaming mode

□□ data packets sent between services (snapshot at a certain point in time)   ▬ service

FIGURE 7.2: Service selection with data streaming tasks pattern.

3. *Data partitioning and batch/streaming pattern* (Figure 7.3)—this case extends the previous one with a possibility of parallelization of data packet processing among parallel paths of a workflow. Tasks working in parallel may process data in batches or in the streaming mode. Routing data packets actually selects workflow paths for processing. Consequently, each task working in parallel with others may have just one service each assigned to it. Some tasks may still require selection of services, especially if working in the batch mode.

4. *Redundancy pattern* (Figure 7.4)—in this case, workflow tasks are generally assigned services to execute the task (making the workflow concrete at this point) but with an additional service per task for using it in case the first service has failed. The proposed algorithms and a workflow management system allow us to handle it transparently to the user.

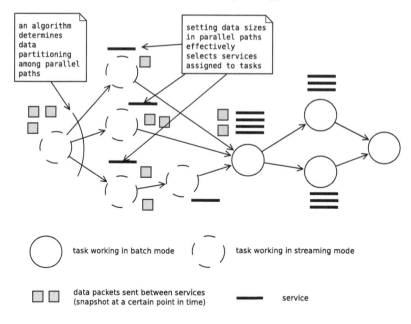

FIGURE 7.3: Data partitioning and batch/streaming pattern.

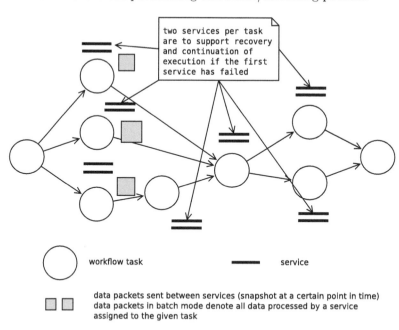

FIGURE 7.4: Redundancy pattern.

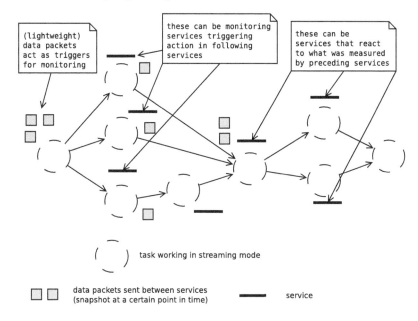

(lightweight) data packets act as triggers for monitoring

these can be monitoring services triggering action in following services

these can be services that react to what was measured by preceding services

( ) task working in streaming mode

data packets sent between services (snapshot at a certain point in time)

service

FIGURE 7.5: Monitoring pattern.

5. *Monitoring pattern* (Figure 7.5)—in this case, input data packets (presumably of small size) are used as triggers in order to invoke following services in a concrete workflow application, e.g., periodically. In fact, the service assigned to the first workflow task may act as a generator that triggers services assigned to following paths by sending data packets. In such a workflow application, services assigned to subsequent tasks may be used for monitoring and reaction to output from monitoring services. Obviously, such a pattern may be combined with, e.g., the *redundancy pattern* for making sure there are alternative monitoring or reaction services in case of failure.

6. *Dynamic changes pattern* (Figure 7.6)—even if some information related to the workflow execution is unknown before the workflow application starts, the models and algorithms can still be used effectively thanks to rescheduling. For instance, if output data size of a service can only be known after execution, workflow rescheduling can be applied at this point. What is more, this allows practical consideration of, e.g., *if* constructs in the workflow. Specifically, after execution of a service assigned to task t_i, only one of several workflow tasks following t_i is activated. Such a case would require re-creation of a part of the graph and subsequent rescheduling.

Obviously, these patterns are given as examples only and many more constructs required by specific applications may be proposed.

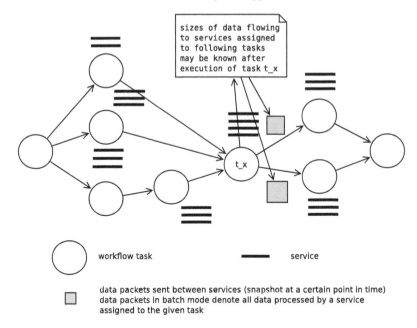

FIGURE 7.6: Dynamic changes pattern.

7.2 Scientific and Multimedia Applications

The workflow examples, solutions, results, and dependencies presented in this section can be used as templates for a variety of practical applications in sciences and multimedia. Scientific and multimedia applications are usually compute-intensive. Some of these process input data of relatively large sizes. Examples of scientific applications include mathematical operations performed on large data sets, simulations of various physical phenomena in the fields of physics, chemistry, and biology, or analysis of data streams from sensors. Multimedia applications usually apply application-oriented algorithms to data such as sound, image, and video, either processed in streams from cameras, microphones, or off-line from collections of stored files. Usually optimization is aimed at decreasing execution time of a time-consuming workflow, maximizing throughput in a streaming mode, or engaging multiple resources such as cluster nodes or clusters in order to cope with large data sizes impossible to handle on a single machine. While performance is a key factor, cost needs to be considered, and power consumption is becoming an important factor. It should be noted that costs can change over time, such as between day and night or between seasons, which suggests optimization of workflow scheduling considering the time of day or year. Other factors such as power consumption of resources on which services were installed can be easily taken in the model presented in Chapter 4. Handling data of large sizes and corresponding costs can be modeled as shown in Chapter 5.

7.2.1 Modeling

In the literature, several workflow application examples were characterized and analyzed thoroughly [122], some of which are:

1. *Epigenomics* [122]—a workflow application with several paths and pipelined computations that is used for mapping the epigenetic state of human cells. The example takes DNA as input. Data is partitioned among parallel paths, processed, merged, and processed until a result is obtained.

2. *CyberShake* [204, 122]—a workflow application for earthquake hazard analysis. The workflow starts with data partitioning for parallel execution of SeismogramSynthesis tasks followed by computation of peak values and saving collected results.

3. *Montage* [187, 204, 195, 122]—an astronomy workflow application for creation of a mosaic of the sky based on input images. The latter are processed in parallel for reprojection, analysis of overlapping, output aggregation, computation of corrections, aggregation into an output, resizing, and format conversion.

4. *Laser Interferometer Gravitational Wave Observatory (LIGO) workflow* [195, 122]—a workflow that splits input from LIGO detectors for subsequent parallel analysis. The workflow contains subworkflows (e.g., up to 100 [122]), which in turn contain parallel paths and aggregation tasks.

Paper [122] gives detailed statistics on running times of each workflow task, I/O operations, CPU, and memory utilization.

Modeling processing as a workflow application according to the model presented in this book allows expression of the following constructs that appear in practical applications:

- Data parallelization using parallel paths—data is either:
    - partitioned among parallel paths a priori, e.g., the particular solution imposes a certain number of partitions and data sizes per tasks are known a priori (solution provided in Section 4.2), or
    - the scheduling algorithm can determine how possibly many data packets can be sent for computations among parallel workflow paths (solution provided in Section 4.4). The latter might correspond to solutions of various execution times and costs. It is possible that such solutions are in fact realized by parallel paths, each of which

is composed of a different number of different tasks. This may be the case if there are various solutions to the same problem.

- Performing various operations on the same/partially overlapping data.

- Synchronization of processing on a task following preceding tasks.

It should be noted that input data to a task can be either processed in a single batch or in a streaming mode in which individual data packets are processed as they arrive and results forwarded to following nodes immediately. Tasks working in various modes can be combined into a single workflow application.

A more detailed discussion on performance and performance-cost constraints dependency for scientific and multimedia workflows will be performed next on the following workflow application examples implemented by the author:

1. Processing multimedia content such as a set of images taken by a digital camera in a RAW format. Parallel paths of the workflow allow for parallel processing of images while particular tasks do have the streaming mode enabled. It should be noted that processing of individual images may require several filters (possibly executed by independent services). Finally, a service such as archiving or Web publishing may be applied to all images. In the latter case, the non-streaming mode must be used. Examples of this type of processing include filters such as RAW- to TIFF-conversion, image normalization, application of custom curves or levels, contrast adjustment, resizing, sharpening [71], etc.

2. Parallel processing of math operations [61] such as matrix operations, computing eigenvectors, and inverse matrices.

The aforementioned examples may be modeled in general by the workflow structure shown in Figure 7.7. The following features can be distinguished for this type of processing:

1. The whole workflow can be divided into subgraphs that partition input data, process, and integrate results.

2. Location of input data is taken into account. It might be important if the input needs to be fetched from a device with a fixed and distant location such as a radiotelescope.

3. It allows for parallel processing of data by independent parallel paths. In each of the paths, there may be a different number of tasks as there may be various solutions to achieve the same goal, e.g., various algorithms for complex matrix operations, etc. Even if services installed in parallel paths use the same resource, the scheduling model takes this fact into account. If the services in distinct paths do not share resources, this makes the scheduling problem easier.

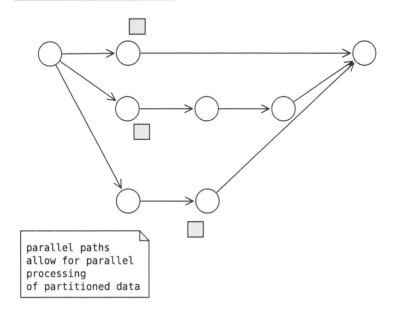

FIGURE 7.7: Scientific/multimedia workflow application for parallel processing.

4. If it is known in advance that some parameters like costs of services will change, e.g., the service cost may be different during the day and night, another section might be added to consider migration of processing to other services, e.g., to another hemisphere. This is depicted in Figure 7.14.

7.2.2 Trade-offs and Performance

Based on real workflow applications analyzed by the author, i.e.:

1. parallel processing of images in [60] and [71] and

2. parallel numerical integration in [61],

it is possible to determine the typical expected execution times and speed-ups for a varying number of paths (and thus resources with services) for such

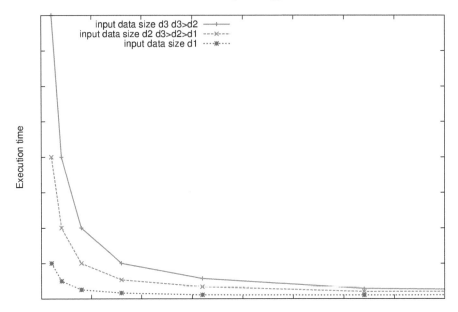

Number of parallel paths with independent services

FIGURE 7.8: Typical graph of workflow execution time vs. the number of paths with independent services.

parallel sections of workflow graphs. These are shown in Figures 7.8 and 7.9, respectively.

Furthermore, as investigated by the author in [71] and [61], there is a visible trade-off related to the size of data packets sent between services in the workflow. The typical function of the total workflow execution time versus the data packet size with the other parameters fixed is shown in Figure 7.10. Specifically:

- For small data packet size the overhead on communication and handling multiple data packets and smaller computations/communication ratio affects performance.

- Larger data packets slow down the start of computations and make it difficult to balance load among parallel workflow paths, especially if the processing paths use services of various performance.

Execution times and speed-up values were analyzed by the author for various

1. total size of input data [60] and

2. sizes of data packets [71].

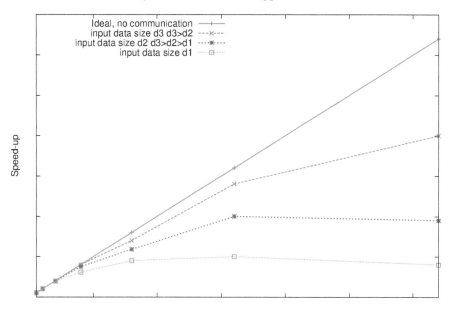

FIGURE 7.9: Typical graph of speed-up vs. the number of paths with independent services.

This is then extended with further experiments considering impact of the cost bound on performance.

Furthermore, the workflow management system proposed in Section 6.2 allows for consideration of alternative computing paths with various quality characteristics that might refer to

1. cost of solutions,

2. reliability, and

3. conformance with requirements, etc.

The proposed model allows for description of services and correspondingly a path of tasks with services using various quality metrics. As an example, Figure 7.11 presents a workflow application with several paths, some of which might be inactive in case of tight budget requirements. Namely, some faster paths were not activated because they were too expensive to be included in the budget. Obviously, this impacts the workflow execution time.

For a real workflow application example, the author used the digital photography workflow from [60] and extended previously obtained results with consideration of budget constraints for various numbers of paths in the workflow. The workflow application uses from 4 to 16 paths with varying costs,

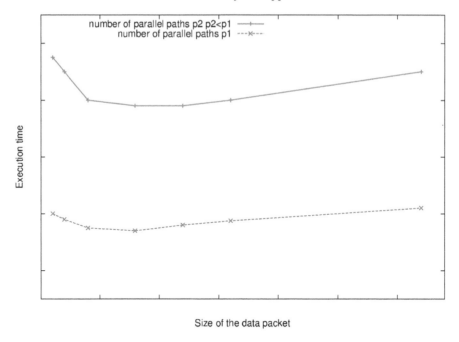

FIGURE 7.10: Typical impact of the data packet size on the workflow execution time.

with 20 per processing time unit for services in 8 paths, 15 per processing time unit for services in 4 paths, and 10 per processing time unit for services in 4 paths. Figure 7.12 shows how the execution time of the workflow application changes with an increasing number of paths and for various constraints on the total cost of selected services. Tighter budgets result in larger execution times. Figure 7.13 shows corresponding speed-up values compared to processing within a single path.

Furthermore, Figure 7.15 extends results shown by the author in [61] for a workflow application for numerical integration of any given function. In this case, relation of the execution time versus budget constraints is very similar.

7.2.3 Checkpoint and Restart for Improving Performance

Scientific and multimedia services tend to be time consuming and because of that are often run on powerful servers or clusters. As explained in Section 4.6, if there are additional applications run in a shared environment, this can change the expected running time of a service. In this case, if this seriously changes the running time of a service, a workflow execution engine may terminate the service and launch rescheduling considering alternative services in order to optimize the goal and still meet imposed constraints. Another solu-

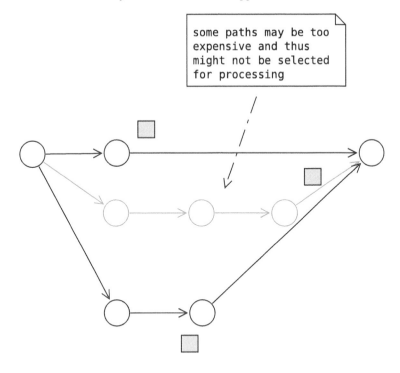

FIGURE 7.11: Scientific/multimedia workflow application with budget constraints.

tion can use a checkpoint and restart in order to move the application behind a service to, e.g., a different node of a cluster. In this case, the running time of a service will only be extended by the checkpoint/restart time (presumably small compared to the service execution time) and may still be within an additional timeout allowed for service running time from the point of view of the workflow execution engine.

The following concrete solution presents the way this can be achieved in a real environment using Berkeley Lab Checkpoint/Restart (BLCR) [177]. BLCR can be used to checkpoint an application running under Linux and restart it afterward, even on a different machine such as a different cluster node. Furthermore, this solution can be used for parallel MPI applications [214].

The script shown in Listing 7.1 wraps an application with an executable specified in **app_name**. Invocation of the script can replace invocation of the application as the script terminates after termination of the application. The script launches the application and monitors loads of cluster nodes. If the load on the node on which the application is running exceeds a threshold, then the application is checkpointed and restarted on a different node of the cluster.

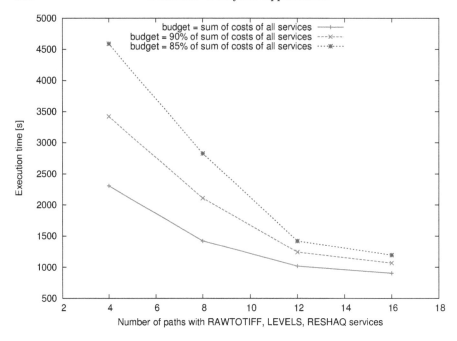

FIGURE 7.12: Execution time vs number of paths with service nodes for various costs for digital image processing workflow application.

This process will continue if loads of the nodes change again. Specifically, the most important steps include:

1. launching an application using `cr_run`—line 12,

2. running a main load monitoring loop (line 55) until the application has terminated, which is checked for in line 59,

3. fetch loads of nodes within a cluster—line 67,

4. check if the load on the current node is too high (line 74) given the number of cores predicted for the application (line 4),

5. check if there are other nodes within the cluster with a lower load—line 80,

6. get a pid of the process to be checkpointed – line 84,

7. checkpointing using `cr_checkpoint --term --save-all`—line 90,

8. restarting the application from a checkpoint file (`context.<pid>`)—line 93, and

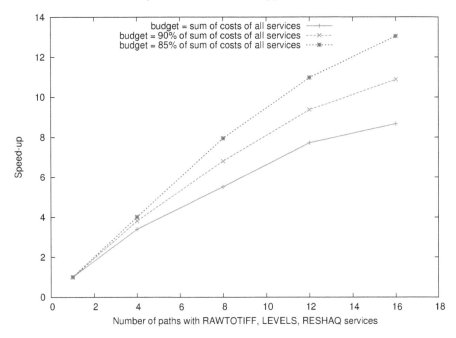

FIGURE 7.13: Speed-up vs number of paths with service nodes for various costs for digital image processing workflow application.

9. storing the number of a new node to which the application has been migrated.

Listing 7.1: Script `manage_run`.

```
app_name="./ ___tt"
current_node_number=0
proc_count='cat /proc/cpuinfo | grep processor | wc -l'
cores_for_app=1
let proc_count_compare=proc_count−$cores_for_app
for i in {0..8}
do
    nodes[$i]=compute−0−$i
done

#start the app on the current node in the background
cr_run $app_name &

get_node_loads(){

for i in {0..8}
do
```

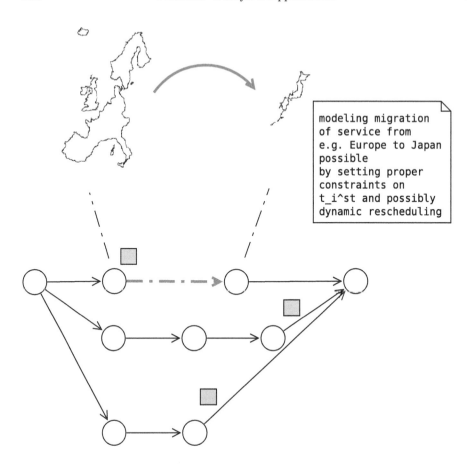

FIGURE 7.14: Scientific/multimedia workflow application for parallel processing with migration of service between continents.

```
18    load[$i]='ssh ${nodes[$i]} uptime | awk '{print $10}' |
          cut -d'.' -f1'
      echo Load of ${nodes[$i]} is ${load[$i]}
    done


    }
23
    get_load_of_current_node(){
      local ret='ssh compute-0-$current_node_number uptime | awk
          '{print $10}' | cut -d'.' -f1'
      #update in the array
      load[$current_node_number]=$ret
28    echo "$ret"
```

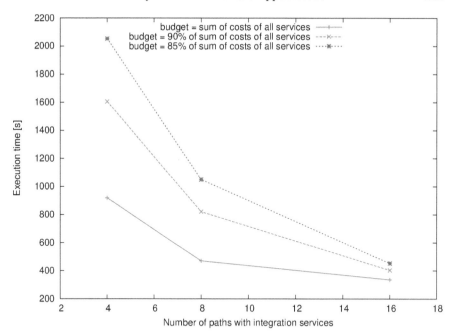

FIGURE 7.15: Execution time vs. number of paths with service nodes for various costs for numerical integration workflow application.

```
}
get_lowest_load_node_number(){
#returns the number of the node with the lowest load
#if the current node has the lowest or equal load — its
    number is returned

local current_node_load=$(get_load_of_current_node)
local ret=current_node_number

for i in {0..8}
do

  if [ $current_node_load −gt ${load[$i]} ]; then
    ret=$i
    current_node_load=${load[$i]}
  fi
done
  echo "$ret"
}
```

```
running=1
sleep_seconds=5
node_load_counter=0
```

53 `get_node_loads`

```
while [ $running != 0 ]; do
  sleep $sleep_seconds
```

58 ```
   #check if the application has terminated
   running=`ssh compute-0-$current_node_number ps aux | grep
     $app_name | head -1 | wc -1`

   let node_load_counter=node_load_counter+1
   #check the load
```
63 ```
   get_loads=$(($node_load_counter % 3))

   if [ $get_loads = 0 ]; then
     echo Getting node loads
     get_node_loads
```
68 ```
   fi

   load_of_current_node=$(get_load_of_current_node)
   load[$current_node_number]=$load_of_current_node
   echo Load of current node [$current_node_number] =
     $load_of_current_node
```
73
```
   if [$load_of_current_node -gt $proc_count_compare]; then
     echo load larger than proc count

     #check if there is a node with lower load
```
78 ```
     lowest_load_node_number=$(get_lowest_load_node_number)

     if [ lowest_load_node_number != current_node_number ];
   then

       #get a pid of the running process in order
```
83 ```
       #to migrate it to another node
       app_pid=`ssh compute-0-$current_node_number ps aux |
     grep $app_name | head -1 | awk '{print $2}'`

       #migrate a process to another node
       echo Migrating $app_name pid=$app_pid from compute
         -0-$current_node_number to compute-0-
         $lowest_load_node_number
```
88
```
       #checkpoint the process on the given node
       ssh compute-0-$current_node_number cr_checkpoint --
   term --save-all $app_pid
```
```

```
        #restart the process on another node
93      ssh compute-0-$lowest_load_node_number cr_restart
     context.$app_pid &

        #change the current node number
        current_node_number=$lowest_load_node_number
     fi
98   fi
  done
```

Listing 7.2 presents a log using script **manage_run** in a real cluster environment. It should be noted that the loads reported in the log are those reported by **uptime** and correspond to other processes running on the nodes. The application used behind a service in this case is for computing an average of 10^k $k \in Z_+$ numbers and in this particular test has an estimated running time of 70 seconds on a single, not loaded node. Initially, the script launches the application on node **compute-0-0** and already sees that the load from other processes would change the application running time compared to the value declared in the service description. Because of that, the script migrates the application to node **compute-0-4**, which is not loaded by other applications (line 12). However, this node becomes overloaded as well by other processes (not considering the application itself) and the script migrates the application to node **compute-0-5** (line 27). For this particular application, the size of the checkpoint file was 2.28 MB with an average time for checkpointing of 0.74 seconds (line 90 in Listing 7.1).

Listing 7.2: Output from script **manage_run**.

```
 1 Load of compute-0-0 is 4
   Load of compute-0-1 is 2
   Load of compute-0-2 is 4
   Load of compute-0-3 is 4
   Load of compute-0-4 is 0
 6 Load of compute-0-5 is 0
   Load of compute-0-6 is 0
   Load of compute-0-7 is 0
   Load of compute-0-8 is 3
   Load of current node [ 0 ] = 4
11 load larger than proc count
   Migrating ./___tt pid=16663 from compute-0-0 to compute-0-4
   ./manage_run: line 105: 16663 Terminated            cr_run
       $app_name
   Load of current node [ 4 ] = 0
   Getting node loads
16 Load of compute-0-0 is 4
   Load of compute-0-1 is 2
   Load of compute-0-2 is 4
   Load of compute-0-3 is 4
```

```
   Load of compute-0-4 is 1
21 Load of compute-0-5 is 0
   Load of compute-0-6 is 0
   Load of compute-0-7 is 0
   Load of compute-0-8 is 3
   Load of current node [ 4 ] = 1
26 load larger than proc count
   Migrating ./___tt pid=16663 from compute-0-4 to compute-0-5
   Load of current node [ 5 ] = 0
   Load of current node [ 5 ] = 0
   Getting node loads
31 Load of compute-0-0 is 4
   Load of compute-0-1 is 2
   Load of compute-0-2 is 4
   Load of compute-0-3 is 4
   Load of compute-0-4 is 2
36 Load of compute-0-5 is 0
   Load of compute-0-6 is 0
   Load of compute-0-7 is 0
   Load of compute-0-8 is 3
   Load of current node [ 5 ] = 0
41 Load of current node [ 5 ] = 0
   Load of current node [ 5 ] = 0
   Average=100.005963 Getting node loads
   Load of compute-0-0 is 4
   Load of compute-0-1 is 2
46 Load of compute-0-2 is 4
   Load of compute-0-3 is 4
   Load of compute-0-4 is 2
   Load of compute-0-5 is 0
   Load of compute-0-6 is 0
51 Load of compute-0-7 is 0
   Load of compute-0-8 is 3
   Load of current node [ 5 ] = 0
```

7.3   Business Production Problem

Business workflow applications are usually more control oriented and process smaller data sets. On the other hand, more QoS parameters are usually considered and changes in service availability and new services showing up are more frequent than for scientific workflows.

It should be noted that in this context a workflow can model one of the following:

- production workflows corresponding to development or manufacturing of products that integrate components produced or delivered by various providers,

- design phases for which the ultimate goal is a design to which various parties contribute with their own services, or

- both of the above modeled in a single workflow application.

In paper [66] the author presented a workflow application for assembly of products out of components provided by third parties and subsequent distribution. Delivery of particular components can be expressed as services with corresponding QoS parameters. The workflow application includes three stages:

1. acquiring components, each modeled as a service and provided by services offered by various providers,

2. integration of the components into a final product,

3. distribution of the product among several markets; similar to the first stage, each market is modeled as a service and distribution is represented by services offered by various providers.

Figure 7.16 presents an extension of that approach toward a nested business workflow example. It allows production of several products at the same time and imposing a single deadline on all of them along with the common budget. This can be more practical than optimization for all individual workflow applications separately.

As a demonstration, the author extended results presented in [66]. For the structure shown in Figure 7.16 within the meta-node using notation proposed in Section 4.3, the parameters for particular workflow tasks and services were as follows:

- purchasing components—execution time/cost: 8/12, 9/10, 10/8, 11/7, 12/6;

- integration—execution time/cost: 8/32;

- distribution of products—execution time/cost: 3/12, 5/6, 7/4.

Figure 7.17 extends the results from [66] by presenting what costs are obtained by an ILP scheduling algorithm (presented in Section 4.2.1) for particular cost bounds and corresponding execution times. It can be seen that the algorithm approaches the bounds closely using the available budget almost to the maximum. This also results in shorter and shorter execution times for larger budgets.

7.4 A Workflow Application for Planning Interdisciplinary Projects

Nowadays, universities, firms, and institutions often deal with others within scopes of interdisciplinary projects. Before such a project starts, project

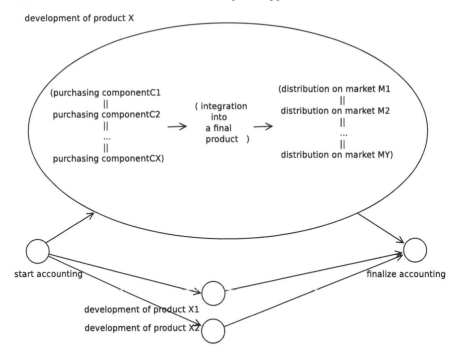

FIGURE 7.16: Nested business workflow example, extending the concept from [66].

preparation takes place. It is usually time consuming and requires much effort related to finding partners who would be able to accomplish parts of the project and actual preparation of the project. The latter includes specification of requirements, definition of acceptance criteria, decomposition into parts, definition of milestones, and preparation of a budget. It should be noted that for such a project, a consortium often needs to involve various types of partners, e.g., firms, universities or research labs, and governmental institutions. Consequently, the process is applicable to many complex projects in various application fields. In case such a consortium of partners applies for external funds, a time frame for project preparation is strictly defined.

Consequently, the model proposed in Chapter 4 allows us to express the following within the scope of this problem:

1. finding partners able to contribute to particular parts of the project,

2. constraints on the cost and reputation/quality of the partners, and

3. definition of a time frame for the project along with milestones and corresponding deadlines.

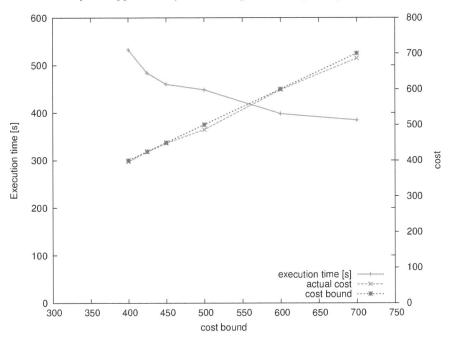

FIGURE 7.17: Execution time vs. cost bound for the business workflow application.

A workflow template that models preparation of a project is shown in Figure 7.18. The following steps are modeled:

1. Project definition including requirement specification and acceptance criteria along with decomposition of the problem into subproblems or parts for which proper partners need to be found.

2. Initial tests, simulations or experiments that need to be performed for some of these parts. These may include assessment of required tools, preliminary time needed for a part or phase, etc.

3. Preparation of actual parts of a project.

4. Global verification of the plan.

5. Corrections based on the verification, preparation of budget estimates, and local finance approval at individual institutions.

6. Synchronization of results.

7. Independent review and final corrections.

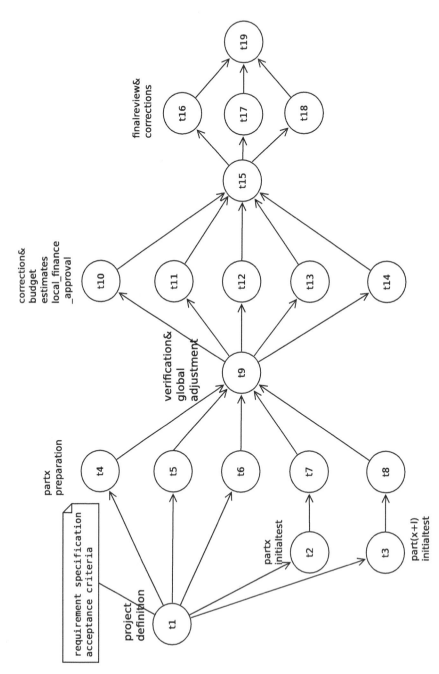

FIGURE 7.18: A workflow application for planning interdisciplinary projects.

Workflow optimization can now be defined as follows. There are several constraints that need to be imposed on the following:

- *Workflow* execution time corresponding to the deadline for project plan preparation i.e., $t_{\text{workflow}} \leq T_{\text{deadline}}$.

- It can be assumed that each party interested in participation in the project would propose their own services for each of relevant tasks, i.e., a j-th provider would propose j-th services for all relevant tasks to a particular part of the project, i.e., assuming sets of tasks $\{t_4, t_{10}\}$, $\{t_5, t_{11}\}$, $\{t_6, t_{12}\}$, $\{t_2, t_7, t_{13}\}$, $\{t_3, t_8, t_{14}\}$ each correspond to one part of the project, $\forall_j$ $P_{4j} = P_{10j}$ $P_{5j} = P_{11j}, P_{6j} = P_{12j}, P_{2j} = P_{7j} = P_{13j}, P_{3j} = P_{8j} = P_{14j}$ following the notation from Section 4.2. Then, selection of a partner for one task would require selection of other services of the same partner for other corresponding tasks. For the integer linear formulation discussed in Section 4.2.1.1 and variables b_{ij}s, this can be forced by imposing $\forall_j$ $b_{4j} = b_{10j}$ $b_{5j} = b_{11j}, b_{6j} = b_{12j}, b_{2j} = b_{7j} = b_{13j}, b_{3j} = b_{8j} = b_{14j}$. Similarly, it can be imposed at the level of the genetic algorithm or GAIN during service substitution.

- *Reputation*—each service s_{ij} might have another parameter assigned to it rp_{ij} where higher values correspond to a higher reputation; it is recommended to obtain high-reputation services and corresponding partners standing behind such services.

In this case, service costs may be understood not as the cost of the project preparation phase, but as expenses predicted by various partners in subsequent project execution. Correspondingly, such real costs may be assigned to services for tasks in the `partx` preparation phase (Figure 7.18). Other services of the parties in the project planning workflow could be either set to 0 or corresponding to other values important during project execution. Given the aforementioned constraints on the workflow running time, a workflow scheduling problem might optimize, e.g., the total expected budget following Equation 2.5, i.e.,

$$min \sum_i c_{i\ sel(i)} d_i \ \text{i.e.,} \ min \sum_{i,j} c_{ij} d_{ij}^{\text{in}}$$

$$t_{\text{workflow}} \leq T_{\text{deadline}}$$
$$\forall_j \ b_{4j} = b_{10j} \ b_{5j} = b_{11j}, b_{6j} = b_{12j}$$
$$b_{2j} = b_{7j} = b_{13j}, b_{3j} = b_{8j} = b_{14j}.$$

Alternatively, there might be an upper bound imposed on the total expenses, e.g., $\sum_i c_{i\ sel(i)} d_i \leq B$, in which case the total reputation of partners might be maximized apart from constraints on individual values, i.e.,

$$max \sum_i rp_{i\ sel(i)} \ \text{i.e.,} \ max \sum_{i,j} rp_{ij}b_{ij}$$

$$\sum_i c_{i\ sel(i)}d_i \leq B$$

$$\forall_j \ b_{4j} = b_{10j} \ b_{5j} = b_{11j}, b_{6j} = b_{12j}$$

$$b_{2j} = b_{7j} = b_{13j}, b_{3j} = b_{8j} = b_{14j}.$$

In the aforementioned formulations, a condition on the minimum reputation for selected services can also be imposed as:

$$\forall_{i,j} rp_{ij} \geq RPMIN \cdot b_{ij}. \tag{7.1}$$

This problem formulation can be easily solved by all the algorithms proposed in Section 4.2.

7.5 Control and Management of Distributed Devices and Sensors

Other than using BeesyCluster and its workflow management system for optimization of service selection and optimization including execution time, cost and possibly other quality metrics such as reliability, the framework can be used as a middleware for integration and control of distributed devices. In this context, the features of BeesyCluster's workflow management system can be used in order to:

1. control remote devices and connecting outputs and inputs of various devices,

2. switch between functionally equivalent services in case of failure,

3. modify the list of available devices quickly by adding and removing these easily.

Workflow management in BeesyCluster allows for very easy integration of various nodes, either general purpose or dedicated to, e.g., high-performance computing. BeesyCluster only assumes invocation of services using SSH and passing inputs as files and gathering outputs as files. A special mode is enabled for nodes with queuing systems such as PBS, LSF, or LoadLeveler in which case invocation is performed using commands specific to the given system.

Another solution must be found for communication of the node accessible from BeesyCluster's workflow management system with external devices. Either a service published in BeesyCluster controls a device on the node on which it is deployed or must use some other communication and control mechanisms to access remote devices.

Practical examples of such applications are described next.

7.5.1 Monitoring Facilities

We assume that there are computers available that have cameras attached to them that gather video streams that are further sent to high-performance computing systems like clusters for further analysis [74]. There are two alternative ways possible for handling the data:

- The computer with the camera may expose a stream, and a service on that computer returns as soon as possible indicating that the stream is ready. BeesyCluster can invoke another service on an HPC node that can fetch the data from the stream.

- The service on the computer equipped with a camera may provide individual frames to subsequent nodes in the workflow graph. In case of high-rate streams and centralized management through BeesyCluster, this may become a bottleneck in processing. Alternatively, another proxy may be set up for communication between services, much like in the case of agents as described in Section 6.2.2.2.

Another example of a system for this type of processing is KASKADA [132, 133]. KASKADA has been specifically designed for processing multimedia streams with specific communication protocols between services of the workflow. This potentially gives higher performance for this type of applications at the cost of flexibility and inability or difficulties in modeling other types of scientific or business applications.

7.5.2 Integration of HPC Nodes and Mobile Devices for Optimization of Traffic and Emergency Procedures

In [54] the author proposed an architecture and a workflow for monitoring and control of a smart home. That solution distinguishes:

- homes that are accessible through services installed on proxy computers;

- mobile devices installed within the homes and using a variety of sensors including light, temperature, noise, accelerators; an advantage of this solution is that it is very easy to change the mobile device location;

- an HPC system that is used for demanding processing of data gathered from the mobile devices, e.g., photos or videos;

- services used for notification of designated forces or institutions such as hospitals, fire stations, or the police.

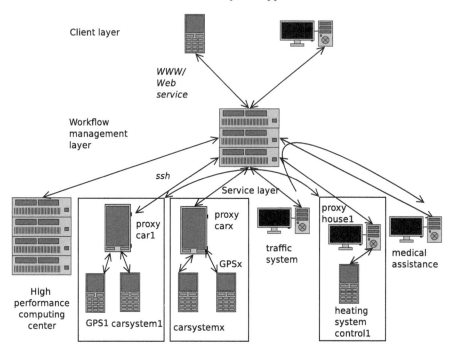

FIGURE 7.19: Architecture for monitoring and optimization of traffic and emergency procedures.

A slightly different approach can be proposed for integration of cars (or other vehicles), medical assistance units, and homes. A corresponding architecture is shown in Figure 7.19 with the following:

- Proxies—tablets located within cars that are connected to the Internet and accessible from a workflow management system.

- mobile devices within the car that report location (GPS) and car parameters.

- An HPC system that is used for demanding processing of data gathered from the mobile devices, e.g., locations of various vehicles and optimizing traffic light control systems.

- Homes with services for controlling home systems such as heating systems.

- Services for notification of medical assistance in case there is an accident involving cars, etc.

The system's operation can be described as two subsystems working independently:

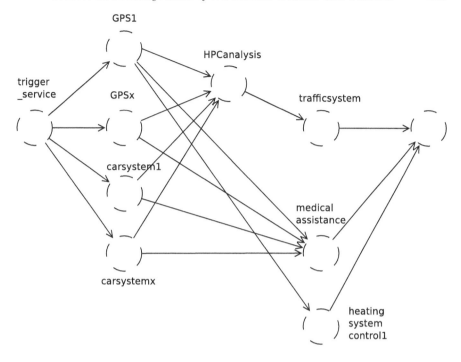

FIGURE 7.20: A workflow application for monitoring and optimization of traffic and emergency procedures.

1. Submission of measurements on possibly various channels to the proxy tablets or servers. This can be done using, e.g., the kSOAP library and invoking Web Services.

2. Execution of the workflow application shown in Figure 7.20. In the workflow, the tasks and corresponding services are executed in the following stages:

   (a) First stage—services are invoked on the proxy servers and the output is passed to either the second or third stage of the workflow application.

   (b) Second stage—services on HPC resources process data efficiently using, e.g., a parallel implementation with MPI. Results may tune a traffic system in the third stage.

   (c) Third stage—services for control of systems or notification of respective forces.

It should be noted that the tasks work in the streaming mode and are triggered by sending data packets periodically in order to activate services and subsequent flows. From this point of view, this workflow application can be *concrete* following Section 2.2.2. On the other hand, if more services for, e.g., medical assistance are assigned, in case of failure, an alternative service will be invoked.

BeesyCluster allows configuration of several independent workflow applications by separate users. Access is granted for particular services and workflow applications to particular BeesyCluster users.

7.5.3 Online Translation on HPC Resources

The proposed architecture for this application is presented in Figure 7.21 and a workflow application in Figure 7.22. The following components are distinguished:

1. Clients who submit recorded samples from mobile devices to a workflow application defined within BeesyCluster. The client may have been granted access to a workflow application with a permission for a certain number of runs. This can be realized by setting the size of input data to the number of runs paid for. In fact, this will correspond to the same number of recorded samples that would be provided to the workflow.

2. Servers or clusters with services installed that are able to translate recorded samples from one language to another. Within the workflow there may be several services differing in execution time and price but also several identical services installed on various nodes. The latter would provide the ability to switch to a new service automatically by the Beesy-Cluster workflow management in case the first one failed [66].

7.5.4 Monitoring of Objects Using Mobile Devices

The application for monitoring of locations of mobile devices presented in Figure 7.24 with the architecture shown in Figure 7.23 has several real-life uses:

1. Monitoring locations of children in dangerous places such as mountains. If a mobile device stays in one spot for too long, it may be because of an accident and trigger verification and automatic notification of emergency.

2. Monitoring of cars in a fleet and, e.g., average, maximum speeds, transfer of goods, etc.

3. Monitoring computers within a distributed company, which can be done for two purposes:

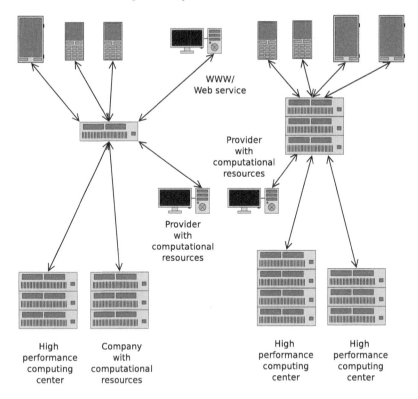

FIGURE 7.21: Architecture for online translation in a distributed environment.

(a) Preventing unauthorized use or theft—computers may report locations either using GPS positioning or their locations can be tracked down using IP addresses and proximities to other servers in the network.

(b) Monitoring usage of resources on the computers, e.g., CPUs, GPUs, disks. If there are several subsidiaries of an international company, this may lead to optimization of assignment of resources to particular subsidiaries. It should be noted that a similar solution was proposed by the author in [62] for volunteer-based computing in the Comcute [6, 75] system. In that case, however, each monitored computer is a client in a volunteer-based system. In this case, each monitored computer has a service assigned to it in one of the ways listed below.

Similar to the online translation example, clients use mobile devices that obtain information about the location of the client. Then the information about the location is sent to a service that gathers subsequent locations of the

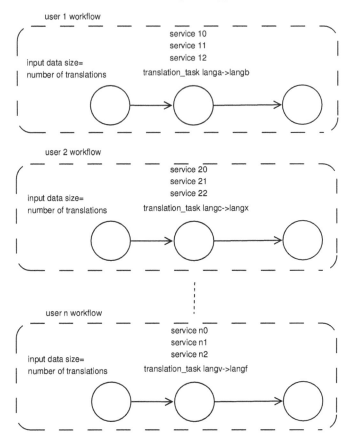

FIGURE 7.22: A workflow application for online translation.

clients and verifies against allowed space for particular clients. Furthermore, for each client there may be a different threshold set that indicates how long the client can stay in the same location before an alarm is triggered. Locations of particular clients can then be viewed by an authorized user in BeesyCluster. Technically, the following configuration is used in BeesyCluster's workflow management system:

1. Services—on mobile devices connected to the Internet with visible IP addresses,

2. Workflow application—separate for each person/a distinct group of people to be monitored.

3. Workflow parameters—in the streaming mode; several input data packets act as triggers to send queries to the mobiles devices, the output indicates the location which can be verified.

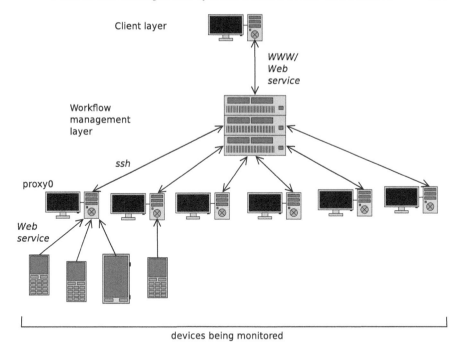

Client layer

WWW/
Web
service

Workflow
management
layer

ssh

proxy0

Web
service

devices being monitored

FIGURE 7.23: Architecture for monitoring of objects in a distributed environment.

4. Privacy—only workflow application owners (e.g., parents) have access to results.

Alternatively, the mobile devices can report their locations periodically:

1. Services—on dedicated computers to which mobile devices connect using, e.g., Web Services and report their locations periodically. The services then would act as proxies that store the data.

2. Workflow application—separate for each person/a distinct group of people to be monitored.

3. Workflow parameters, streaming—several input data packets act as triggers to send queries to the services; output indicates the location which can be verified; however, the service needs to store not only the location but also the last time a mobile device reported to distinguish between not reporting and not moving devices.

4. Privacy—only workflow application owners have access to results.

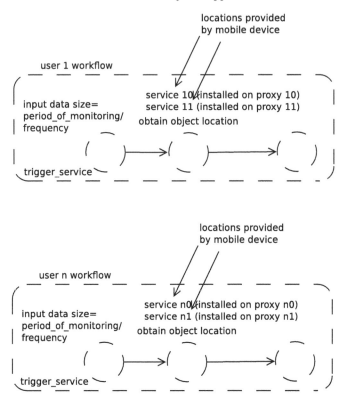

FIGURE 7.24: A workflow application for object monitoring.

7.6   A Script-Based Workflow Application for Parallel Data Processing

In this case, a workflow application for parallel processing data sets is proposed by the author. Particular services assigned to tasks of the workflow application are implemented using bash scripts.

Big Data [153, 33] has recently gathered much attention due to potentially many applications in the modern world.

Since data is essentially stored in files, the latter can be divided into sets on which needed operations and analysis can be performed in parallel. However, there are a few factors that make this problem difficult and different from typical embarrassingly parallel problems [218]:

1. Depending on the analysis algorithm, processing of data fetched from the disk or a disk array is fast compared to data reading. With an increasing

number of processes and threads that read the data in parallel, the storage becomes a bottleneck in the application.

2. Fetching the data from a distant location may take much time. The solution might process already downloaded data files while new ones are being downloaded.

The author implemented a general solution that is based on Unix scripts that can be published as BeesyCluster services. The scripts use the widely available Unix commands for processing files and file content while providing a way to define processing commands or applications in one of the scripts. This is in contrast to the implementation of services in a similar workflow using parallel multithreaded applications that was proposed by the author in [67].

For test purposes in this book, the following scripts were implemented:

- `process_data_packet`—processes the given data packet with an optional argument (Listing 7.3).

- `process_data_packet_multiple_inputs`—launches processing of the given data packet with several, potentially various, inputs (Listing 7.4). For each option (which can be understood as a different input file) a separate application instance can be launched. Such instances may be executed in parallel.

- `process_data_packet_multiple_options`—this script invokes processing of data packets with possibly various options for each spawned instance (Listing 7.5). This is useful if the node contains multiple cores that may process independent data in parallel. On the other hand, the script allows you to control the level of parallelism by waiting for completion of spawned processes in stages. There is a trade-off here between the level of parallelism (particular scripts may execute very quickly) and potential overhead on switching between scripts working on separate files.

- `data_split`—a script that divides the contents of the directory into lists of files to be processed in parallel (Listing 7.6). Such files form data packets to be processed in parallel in the next stages. Note that the script partitions data based on the number of files rather than sizes. This is to speed up partitioning. This strategy may give reasonable partitioning much quicker than when considering file sizes in case of a very long list of files.

Parallel processing can be launched in one of two alternative ways:

1. Within BeesyCluster with the workflow application graph shown in Figure 7.25a. Upon partitioning of the initial directory with its content

into lists of files, the files are then given as input for parallel processing by distinct nodes of a cluster. BeesyCluster offers the ability to restart processing on other nodes in case some services have failed. This implementation is sufficient if partitioning is a relatively quick process. Otherwise, it is possible to use the streaming mode for configuration of pipelining in the following way (Figure 7.25b):

(a) The service based on `data_split` would run in the streaming mode as would the one based on `process_data_packet_multiple_options`.

(b) `data_split` would partition into a larger number of smaller files (with a smaller number of files in the directory to be analyzed). As soon as a small list is available for analysis, it is forwarded to one of the following `process_data_packet_multiple_options` services running on cluster nodes.

2. Using script `parallel_do` (Listing 7.7), which is a top-level script that launches the same process BeesyCluster would do. The script would entail a slightly smaller overhead due to verification of the user credentials and communication with BeesyCluster with either WWW or Web Services (both over HTTPS). Launching processing is performed using `ssh` in both cases. The workflow graph can be modeled as shown in Figure 7.26 with tasks marked as working either in a streaming or batch processing mode.

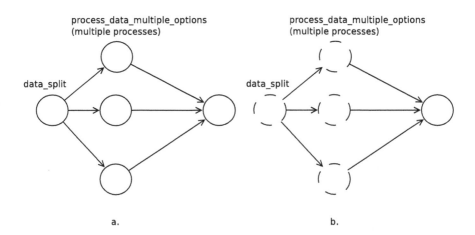

FIGURE 7.25: A workflow application for parallel data analysis with services deployed in BeesyCluster.

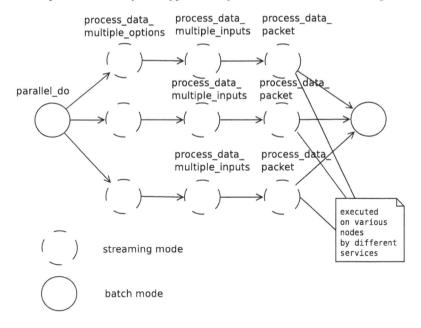

FIGURE 7.26: A workflow application for parallel data analysis; a logical graph in which services are realized by scripts.

Furthermore, and very important, the parallelization level of the analysis section of the workflow can be controlled in the following way:

- We assume that distinct services are defined in such a way that each runs on a separate node, e.g., of a cluster node. In line with the model presented in Section 5.4.3, the maximum size of data processed by each service can be set to such a value that the number of files processed does not exceed the local storage if the latter is used in analysis.

- Additionally, the nodes may use file systems such as Lustre [192]. One may need to restrict the *total* number of parallel services that refer to the storage. This can be set in BeesyCluster, which monitors the storage used by individual services. In fact, this is related to the total workflow data footprint, which is the total data size used by the workflow application at the given moment (Section 5.4.5). In this application, the data size is expressed by the number of data files, each of which will be handled by a separate service/process.

Listing 7.3: Script `process_data_packet`

```
#process data packet $2 with input $1 and save occurrence in
    file $3
2 process_data_packet $1 $2 >> $3.'hostname'
```

Listing 7.4: Script `process_data_packet_multiple_inputs`.

```
# format <option> <data_packet>
    grep $1 $2 > tempout
3   tempfilesize=$( stat -c%s "tempout")
    if [ $tempfilesize -gt 0 ] ; then
        # now invoke processing for a given data packet with
            possibly various inputs

        ./process_data_packet <input1> $2   <filex >.$1 &
8       ./process_data_packet <input1> $2   <filey >.$1 &
        ...
        ./process_data_packet <input2> $2   <filex >.$1 &
        ./process_data_packet <input2> $2   <filey >.$1 &
    fi
```

Listing 7.5: Script `process_data_packet_multiple_options`.

```
# parameters <file with data packet list> <file_with_options
    >
# the solution is as follows: read the file line by line ,
    each line denotes a file
3 # for each of the files do the following

tr ' ' '\n' < $2 > $2.options
i="0"
cat $2.options | while read option; do
8 cat $1 | while read line; do
    ./process\_data\_packet\_multiple\_inputs $option $line &
    # wait if too many invocations
    if [ $i -eq 8 ]
    then
13    i="0"
    wait
    fi
    i=$[ $i+1]
done
18 done
```

Listing 7.6: Script `data_split`.

```
#arguments:
2 #directory that should be split
#the number of cluster nodes - files to which the directory
    should be split
rm input.compute-0-*
find $1 -type f > files
totalfiles='wc -l files | cut -f1 -d' ' '
7 files0=$(( totalfiles / $2 ))

split -a 1 -d -l $files0 files input.compute-0-
```

```
   # this has produced files such as input.compute−0−1
12 # we need to add the contents of input.compute−0−$2 to input
      .compute−0−0
   cat input.compute−0−$2 >> input.compute−0−0
   rm input.compute−0−$2
```

Listing 7.7: Script `parallel_do`.

```
 1 #this is the top level script that will take
   # 1. an input − a directory
   # 2. and a file with options
   # 3. the number of cluster nodes used
   # 4. the number of cores per node used
 6
   # determine the prefix used for running on a certain number
      of cores
   case $4 in
         1)
                 coretext=" taskset −c 0 "
11               ;;
         2)
                 coretext=" taskset −c 0,1 "
                 ;;
         4)
16               coretext=" taskset −c 0−3 "
                 ;;
         8)
                 coretext=" "
                 ;;
21       *)
                 echo "Allowed number of cores is 1/2/4/8
                    "
                 exit
                 ;;
   esac
26
   # partition the files for parallel processing by separate
      cluster nodes
   ./data_split $1 $3

   # and start processing on the desired number of cluster
      nodes!
31 # do it in a loop
   i="0"
   while [ $i −lt $3 ]
   do
   proc=$i
36 ssh compute−0−$proc "cd code ; $coretext ./ process \ data \
      packet \ multiple \ options ./ input.compute−0−$i $2" &
```

```
i=$[ $i +1]
done

wait
```

The following tests were conducted on a cluster with 10 nodes, each with 2 dual-core processors and HT technology for a total of 8 logical cores per node. The specifications of each node are Intel Xeon CPU 2.80 GHz, 4 GB RAM, Linux version 2.6.32-220.13.1.el6.x86_64. The connection links used Gigabit Ethernet. The cluster uses a Dell EMC AX150 disk array connected to one of the nodes.

The application used in this case processed a set of input files initially stored in a single directory. The testbed case used 64 large files, each 131 MB in size. The application's goal was the following:

1. Search for an occurrence of a given keyword, implemented using **grep** in Listing 7.3.

2. If found, search for other parameters that may describe the previously found keyword. Practically, it was implemented within Listing 7.4 by looking for occurrence of words associated with the given keyword in the file.

There are several potential applications of this code, like searching for occurrence of specific keywords in books, marketing, security, databases of signals from radiotelescopes, verification of claims in the number theory, etc.

Figure 7.27 presents the execution times while Figure 7.28 presents the obtained speed-ups for this application, the latter compared to a single execution core. The following aspects should be taken into account when analyzing the speed-up:

1. Depending on the file, whether the given keyword is found or not, this may introduce imbalance in the analysis times of various files and potentially impact the speed-up.

2. Parallel access to files (although done independently for the files) is limited by the hardware solutions in the disk array and the latency and bandwidth of the network.

3. The cluster features processing cores, 4 per node, for a total of 32 cores and the HyperThreading technology. While the latter allows us to see 8 logical processors per node for a total of 64 logical processors, the performance is naturally worse than 64 physical cores.

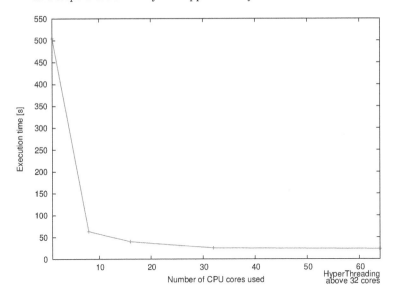

FIGURE 7.27: Execution time vs. number of CPUs and cores for parallel processing of data.

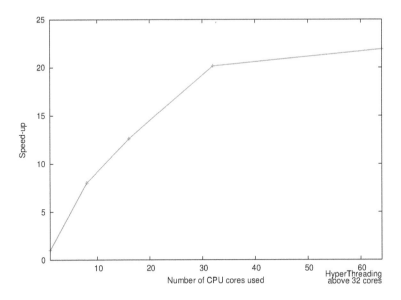

FIGURE 7.28: Speed-up vs. number of CPUs and cores for parallel processing of data.

7.7 Modeling Performance in BeesyCluster

7.7.1 Overhead of the BeesyCluster Workflow Solution

When deployed in BeesyCluster, all the presented workflow applications can benefit from the following features:

1. Automatic reselection of a service for a workflow task after service execution was unsuccessful. This may be the case due to one of several reasons:

    - The service has failed—this requires setting a timeout after which the service is assumed to be unavailable.

    - The network connection has failed/the service cannot be reached— requires setting a maximum number of retries before BeesyCluster assumes the process has failed.

2. Learning about the environment. In case of failure, another service can be chosen, either on the same node (the first case) or on another node (the second case). Furthermore, the system can learn at runtime about the following:

    (a) access to particular nodes on which services were installed,

    (b) failure of particular services—a service may fail for a specific set of input data but work flawlessly for other data sets.

3. Automatic partitioning of data into parallel streams in a workflow as described in Section 4.4.

4. Automatic handling of storage constraints for the services and the whole workflow application as described in Section 5.1.

Naturally, these advantages come at the cost of additional overhead that should be measured and evaluated.

In [66] the author showed the overhead of the BeesyCluster workflow management system when running services installed on nodes of a cluster with Intel Xeon 2.8-GHz processors. Specifically, a workflow application with a certain number of parallel paths each of which has a predefined length was run. The workflow used services getting the local time. The services were activated one after another according to the workflow graph with several parallel paths being executed at the same time. This shows the overhead for initialization such as checking input data (browsing potentially a few locations for determination of data sizes, 10 locations in the experiment), optimization, control, and synchronization of workflow tasks in BeesyCluster in a distributed setting with data of minimal size (a few bytes). For a given number of parallel paths p and length in tasks l, the overhead can be modeled as a linear function $t_o(p) = \alpha_o + \beta_o l$. For $p = 1$, linear regression gives $\alpha_o = 4.5$ [s], $\beta_o = 2.6$ [s] (correlation coefficient 0.99999), $p = 2$, linear regression gives $\alpha_o = 4.64$ [s], $\beta_o = 2.59$ [s] (correlation coefficient 0.99998), $p = 4$, linear regression gives $\alpha_o = 4.88$ [s], $\beta_o = 2.61$ [s] (correlation coefficient 0.99983).

However, a workflow application will usually be naturally distributed, i.e., will need to use distributed services (either for parallelization or because these are only available from other locations), and it is desirable to compare the overhead of parallel execution in BeesyCluster to another, lower-level parallel implementation. In this case an MPI implementation run on two clusters can be compared to a parallel workflow in BeesyCluster in the same environment. Naturally, MPI does not support the BeesyCluster features mentioned above. The overhead of BeesyCluster for the parallel integration example compared to the native MPI solution is presented in [61]. The overhead of the solution to pure MPI is clearly seen for a small data set but becomes reasonable for larger data sets for which real parallel applications are designed. The BeesyCluster workflow achieves the speed-up of 11 on a total of 16 nodes on 2 clusters (compared to 1 node, with a drop of 0.95 compared to pure MPI) for a test case running around 300 seconds.

7.7.2   Modeling Performance of Applications

Based on the parallel multimedia application presented in Section 7.2.2, the author matched parameters of the model against real execution times in order to verify that there are no unexpected factors in processing. The execution of the application in the non-streaming mode can be modeled as follows [61]:

$$t_{exec}(A, d, p) = \alpha + \beta \frac{d}{p} + \gamma d \qquad (7.2)$$

where A denotes the application, d the size of the input data, p the number of parallel paths of the workflow application, and α, β, and γ are coefficients. Component $\beta \frac{d}{p}$ corresponds to processing while γd corresponds to communication.

Compared to [61], this section contains detailed results from the regression that are shown in Figure 7.29. The following reasoning can be applied:

$$t_{exec}(A, 2d, 16) - t_{exec}(A, 2d, 8) = \beta(\frac{2d}{16} - \frac{d}{8}) + \gamma(2d - d) = \gamma d. \qquad (7.3)$$

After γ has been determined, Figure 7.29 can be drawn where $t_{exec}(A, d, p) - \gamma d$ is shown against $\frac{d}{p}$. The figure shows results for $p = \{1, 4, 8, 12, 16\}$ parallel

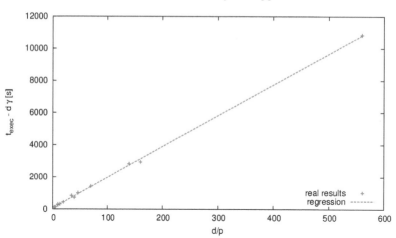

FIGURE 7.29: Modeling execution time of applications in BeesyCluster; linear regression for parallel processing of images.

paths in a balanced workflow application for processing of $d = \{40, 160, 560\}$ digital images. Linear regression is shown. Extending the above comparison, it is possible to model the numerical integration example from [61] in the same way. The results are shown in Figures 7.30 and 7.31.

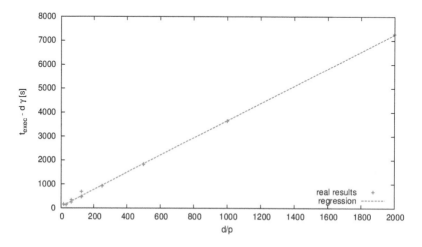

FIGURE 7.30: Modeling execution time of applications in BeesyCluster; linear regression for numerical integration.

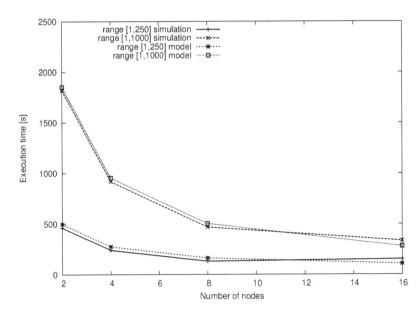

FIGURE 7.31: Modeling execution time of applications in BeesyCluster; comparison of the model and results for numerical integration.

Bibliography

[1] Activiti. http://www.activiti.org/.

[2] Apache ode. the orchestration director engine executes business processes written following the ws-bpel standard. http://ode.apache.org/index.html.

[3] AudioTag. http://audiotag.info.

[4] BeesyCluster Website. https://beesycluster.eti.pg.gda.pl:10030/ek/AS_LogIn.

[5] Cloud computing price comparison engine. http://www.cloudorado.com/.

[6] Comcute Website. http://comcute.eti.pg.gda.pl/.

[7] Globus Toolkit. http://www.globus.org.

[8] IIOP.NET. http://iiop-net.sourceforge.net/.

[9] Introduction to lp_solve 5.5.2.0. http://lpsolve.sourceforge.net/5.5.

[10] jBPM. http://www.jbpm.org/.

[11] Kepler. https://kepler-project.org/.

[12] MigratingDesktop. http://desktop.psnc.pl/pages.php?content=page_main.

[13] omniORB : Free CORBA ORB. http://omniorb.sourceforge.net/.

[14] Openstack. http://www.openstack.org/.

[15] ORBit2. http://orbit-resource.sourceforge.net/.

[16] Orchestra. http://orchestra.ow2.org/xwiki/bin/view/Main/WebHome.

[17] rCUDA. remote CUDA. http://www.rcuda.net/.

[18] Taverna. http://www.taverna.org.uk/.

[19] H. Meuer, E. Strohmaier, J. Dongarra, H. Simon, and M. Meyer. Top500 supercomputer sites. http://top500.org/.

[20] Triana. http://www.trianacode.org/.

[21] *UDDI Version 3.0.2, UDDI Spec Technical Committee Draft.* Editors: L. Clement, A. Hately, C. von Riegen, T. Rogers, http://www.uddi.org/pubs/uddi_v3.htm.

[22] UNICORE. http://www.unicore.eu/.

[23] VBOrb. http://www.martin-both.de/vborb.html.

[24] SHaring Interoperable Workflows for large-scale scientific simulations on Available DCIs (SHIWA), 2010. Grant from the European Commission's FP7 INFRASTRUCTURES-2010-2 call under grant agreement no. 261585, provided by the MTA SZTAKI Computer and Automation Research Institute—Maintained by Laboratory of Parallel and Distributed Computing with Free and Open Source Software, http://www.shiwa-workflow.eu/.

[25] NEARBUY Launches Its Unique Shopping Location Based Mobile App in Dubai, July 2013. ME NewsWire, http://me-newswire.net/news/7838/en.

[26] M. Abouelhoda, S. Issa, and M. Ghanem. Tavaxy: Integrating Taverna and Galaxy workflows with cloud computing support. *BMC Bioinformatics*, 13(1), 2012.

[27] R. Aggarwal, K. Verma, J. Miller, and W. Milnor. Constraint Driven Web Service Composition in METEOR-S. In *Proceedings of IEEE International Conference on Services Computing (SCC'04)*, pages 23–30, 2004.

[28] R. Aggarwal, K. Verma, J. Miller, and W. Milnor. Dynamic Web Service Composition in METEOR-S. Technical report, LSDIS Lab, Computer Science Dept., UGA, May 2004.

[29] S.A. Ahson and M. Ilyas, editors. *Cloud Computing and Software Services: Theory and Techniques.* CRC Press, 2011. ISBN 978-1-4398-0315-8.

[30] D.P. Anderson. BOINC: A System for Public-Resource Computing and Storage. In *Proceedings of 5th IEEE/ACM International Workshop on Grid Computing*, Pittsburgh, USA, November 2004.

[31] S. Artemio, B. Leonardo, J. Hugo, M.J. Carlos, M.A. Aceves-Fernandez, and P.J. Carlos. Evaluation of CORBA and Web Services in Distributed Applications. In *Electrical Communications and Computers (CONIELECOMP), 2012 22nd International Conference on*, pages 97–100, 2012.

[32] B. Aspvall and R.E. Stone. Khachiyan's Linear Programming Algorithm. *Journal of Algorithms*, 1(1):1–13, 1980.

[33] B. Baesens. *Analytics in a Big Data World: The Essential Guide to Data Science and Its Applications*. Wiley, 2014. ISBN: 978-1118892701.

[34] A. Bala and I. Chana. A Survey of Various Workflow Scheduling Algorithms in Cloud Environment. In *IJCA Proceedings on 2nd National Conference on Information and Communication Technology NCICT*, pages 26–30, New York, USA, 2011. Foundation of Computer Science.

[35] P. Bala, K. Baldridge, B. Schuller, N. Williams, E. Benfenati, M. Casalegno, U. Maran, L. Miroslaw, V. Ostropytskyy, K. Rasch, S. Sild, and R. Schne. UNICORE: a Middleware for Life Sciences Grids. In *Handbook of Research on Computational Grid Technologies for Life Sciences, Biomedicine and Healthcare*, ed.: M. Cannataro, IGI Global, Hershey, USA, 2009. - 978-1-60566-374-6. - S. 615 - 643, 2009. Record converted from VDB: 12.11.2012.

[36] P. Balaji, W. Bland, W. Gropp, R. Latham, H. Lu, A. Pena, K. Raffenetti, S. Seo, R. Thakur, and J. Zhang. *MPICH User's Guide*. Mathematics and Computer Science Division, Argonne National Laboratory, October 2014. `http://www.mpich.org/static/downloads/3.1.3/mpich-3.1.3-userguide.pdf`.

[37] J. Balicki, H. Krawczyk, and E. Nawarecki, editors. *Grid and Volunteer Computing*. Gdansk University of Technology, Faculty of Electronics, Telecommunication and Informatics Press, Gdansk, 2012. ISBN: 978-83-60779-17-0.

[38] R. Banger and K. Bhattacharyya. *OpenCL Programming by Example*. Packt Publishing, December 2013. ISBN: 978-1849692342.

[39] A. Barak, T. Ben-nun, E. Levy, and A. Shiloh. A Package for OpenCL Based Heterogeneous Computing on Clusters with Many GPU Devices. In *Proc. of Int. Conf. on Cluster Computing*, pages 1–7, September 2011.

[40] J. Barr. Amazon Web Services Blog. Amazon Simple Workflow—Cloud-Based Workflow Management, Fabruary 2012. `http://aws.amazon.com/blogs/aws/amazon-simple-workflow-cloud-based-workflow-management/`.

[41] S. Beco, B. Cantalupo, L. Giammarino, N. Matskanis, and M. Surridge. OWL-WS: A Workflow Ontology for Dynamic Grid Service Composition. In *e-Science and Grid Computing, 2005. First International Conference on*, page 8, 2005.

[42] F.L. Bellifemine, G. Caire, and D. Greenwood. *Developing Multi-Agent Systems with JADE*. Wiley, 2007. ISBN-13: 978-0470057476.

[43] K. Benedyczak, M. Stolarek, R. Rowicki, R. Kluszczynski, M. Borcz, G. Marczak, M. Filocha, and P. Bala. Seamless Access to the PL-Grid e-Infrastructure Using UNICORE Middleware. In M. Bubak, T. Szepieniec, and K. Wiatr, editors, *PL-Grid*, volume 7136 of *Lecture Notes in Computer Science*, pages 56–72. Springer, 2012.

[44] C. Bexelius, M. Lf, S. Sandin, Y. Trolle Lagerros, E. Forsum, and J.E. Litton. Measures of Physical Activity Using Cell Phones: Validation Using Criterion Methods. *J Med Internet Res*, 12(1), 2010. doi: 10.2196/jmir.1298, PMID: 20118036.

[45] R. Bolze and E. Deelman. *Cloud Computing and Software Services: Theory and Techniques*, chapter Exploiting the Cloud of Computing Environments: An Application's Perspective. CRC Press. Taylor & Francis Group, 2011. ISBN 978-1-4398-0315-8.

[46] R. Buyya, editor. *High Performance Cluster Computing, Architectures and Systems*. Prentice Hall, 1999.

[47] R. Buyya, editor. *High Performance Cluster Computing, Programming and Applications*. Prentice Hall, 1999.

[48] G. Canfora, M. Di Penta, R. Esposito, and M.L. Villani. A Lightweight Approach for QoS-Aware Service Composition. *In Proc. 2nd International Conference on Service Oriented Computing (ICSOC04)*, short papers.

[49] B. Chapman, G. Jost, and R. van der Pas. *Using OpenMP: Portable Shared Memory Parallel Programming (Scientific and Engineering Computation)*. The MIT Press, 2007.

[50] Dave Coffin. Decoding RAW Digital Photos in Linux. `http://www.cybercom.net/~dcoffin/dcraw/`.

[51] PL-Grid Consortium. PL-Grid. `http://www.plgrid.pl/en`.

[52] R. Cushing, G.H.H. Putra, S. Koulouzis, A. Belloum, M. Bubak, and C. de Laat. Distributed Computing on an Ensemble of Browsers. *Internet Computing, IEEE*, 17(5):54–61, 2013.

[53] K. Czajkowski, D. Ferguson, I. Foster, J. Frey, S. Graham, I. Sedukhin, D. Snelling, S. Tuecke, and W. Vambenepe. The WS-Resource Framework, March 2004. `http://www.globus.org/wsrf/specs/ws-wsrf.pdf`.

[54] P. Czamul. Design of a Distributed System Using Mobile Devices and Workflow Management for Measurement and Control of a Smart Home and Health. In *Human System Interaction (HSI), 2013 6th International Conference on*, pages 184–192, 2013.

[55] P. Czarnul. Architecture and Implementation of Distributed Data Storage using Web Services, CORBA and PVM. In Wyrzykowski, R., Dongarra, J., Paprzycki, M., Wasniewski, J., editors, *Parallel Processing and Applied Mathematics*, volume 3019 of *Lecture Notes in Computer Science*, pages 360–367, 2003. 5th International Conference on Parallel Processing and Applied Mathematics, Czestochowa, Poland, Sep 07–10, 2003.

[56] P. Czarnul. PVMWebCluster: Integration of PVM Clusters using Web Services and CORBA. In Dongarra, J., Laforenza, D., Orlando, S., editors, *Recent Advances In Parallel Virtual Machine And Message Passing Interface*, volume 2840 of *Lecture Notes in Computer Science*, pages 268–275, 2003. 10th European Parallel Virtual Machine and Message Passing Interface Users Group Meeting (PVM/MPI), Venice, Italy, Sep 29–Oct 02, 2003.

[57] P. Czarnul. Integration of Compute-Intensive Tasks into Scientific Workflows in BeesyCluster. In Alexandrov, V.N. and VanAlbada, G.D. and Sloot, P. and Dongarra, J., editor, *Computational Science - ICCS 2006, PT 3, Proceedings*, volume 3993 of *Lecture Notes in Computer Science*, pages 944–947, 2006. 6th International Conference on Computational Science (ICCS 2006), Reading, England, May 28-31, 2006.

[58] P. Czarnul. Reaching and Maintaining High Quality of Distributed J2EE Applications: BeesyCluster Case Study. In Sacha, K, editor, *Software Engineering Techniques: Design for Quality*, volume 227 of *International Federation for Information Processing*, pages 179–190, 233 Spring Street, New York, NY 10013, United States, 2006. IFIP, Springer. IFIP Working Conference on Software Engineering Techniques, Warsaw, Poland, Oct 17–20, 2006.

[59] P. Czarnul. BC-MPI: Running an MPI Application on Multiple Clusters with BeesyCluster Connectivity. In *Proceedings of Parallel Processing and Applied Mathematics 2007 Conference,*. Springer Verlag, May 2008. Lecture Notes in Computer Science, LNCS 4967.

[60] P. Czarnul. Modelling, Optimization and Execution of Workflow Applications with Data Distribution, Service Selection and Budget Constraints in BeesyCluster. In *Proceedings of 6th Workshop on Large Scale Computations on Grids and 1st Workshop on Scalable Computing in Distributed Systems, International Multiconference on Computer Science and Information Technology*, pages 629–636, October 2010. IEEE Catalog Number CFP0964E.

[61] P. Czarnul. Parallelization of Compute Intensive Applications into Workflows Based on Services in BeesyCluster. *Scalable Computing: Practice and Experience*, 12(2):227–238, 2011.

[62] P. Czarnul. *Grid and Volunteer Computing*, chapter On Configurability of Distributed Volunteer-Based Computing in the Comcute System, pages 53–69. Gdansk University of Technology, 2012. ISBN 978-83-60779-17-0.

[63] P. Czarnul. Integration of Cloud-Based Services into Distributed Workflow Systems: Challenges and Solutions. *Scalable Computing: Practice and Experience*, 13(4):325–338, 2012. ISSN 1895-1767.

[64] P. Czarnul. A Model, Design and Implementation of an Efficient Multithreaded Workflow Execution Engine with Data Streaming, Caching, and Storage Constraints. *Journal of Supercomputing*, 63(3):919–945, MAR 2013.

[65] P. Czarnul. An Evaluation Engine for Dynamic Ranking of Cloud Providers. *Informatica, An International Journal of Computing and Informatics*, 37(2):123–130, 2013. ISSN 0350-5596.

[66] P. Czarnul. Modeling, Run-Time Optimization and Execution of Distributed Workflow Applications in the Jee-Based Beesycluster Environment. *Journal of Supercomputing*, 63(1):46–71, Jan 2013.

[67] P. Czarnul. A Workflow Application for Parallel Processing of Big Data from an Internet Portal. In *Proceedings of the International Conference on Computational Science, Procedia Computer Science*, Cairns, Australia, June 2014. In press.

[68] P. Czarnul. Comparison of Selected Algorithms for Scheduling Workflow Applications with Dynamically Changing Service Availability. *Journal of Zhejiang University SCIENCE C*, 15(6):401–422, 2014.

[69] P. Czarnul, M. Bajor, M. Fraczak, A. Banaszczyk, M. Fiszer, and K. Ramczykowska. Remote Task Submission and Publishing in BeesyCluster: Security and Efficiency of Web Service Interface. In Springer-Verlag, editor, *Proc. of PPAM 2005*, volume LNCS 3911, pages 220–227, Poland, Sept. 2005.

[70] P. Czarnul, A. Ciereszko, and M. Fraczak. Towards Efficient Parallel Image Processing on Cluster Grids Using GIMP. In Bubak, M., VanAlbada, G.D., Sloot, P.M.A., Dongarra, J.J., editors, *Computational Science - ICCS 2004, Pt 2, Proceedings*, volume 3037 of *Lecture Notes in Computer Science*, pages 451–458, 2004. 4th International Conference on Computational Science (ICCS 2004), Krakow, Poland, Jun 06–09, 2004.

[71] P. Czarnul, T. Dziubich, and H. Krawczyk. Evaluation of Multimedia Applications in a Cluster Oriented Environment. *Metrology and Measurement Systems*, XIX(2):177–190, 2012.

[72] P. Czarnul and M. Fraczak. New User-Guided and Ckpt-Based Checkpointing Libraries for Parallel MPI Applications. In DiMartino, B., Kranzlmuller, D., and Dongarra, J., editors, *Recent Advances in Parallel Virtual Machine and Message Passing Interface, Proceedings*, volume 3666 of *Lecture Notes in Computer Science*, pages 351–358, 2005. 12th European Parallel-Virtual-Machine-and-Message-Passing-Interface-Users-Group Meeting (PVM/MPI), Sorrento, Italy, Sep 18–21, 2005.

[73] P. Czarnul and K. Grzeda. Parallel Simulations of Electrophysiological Phenomena in Myocardium on Large 32 and 64-bit Linux Clusters. In Kranzlmuller, D., Kacsuk, P., and Dongarra, J., editors, *Recent Advances In Parallel Virtual Machine and Message Passing Interface, Proceedings*, volume 3241 of *Lecture Notes In Computer Science*, pages 234–241, 2004. 11th European Parallel Virtual Machine and Message Passing Interface Users Group Meeting, Budapest, Hungary, Sep 19–22, 2004.

[74] P. Czarnul and W. Kicior. Workflow Application for Detection of Unwanted Events. In *Information Technology (ICIT), 2010 2nd International Conference on*, pages 25–28, 2010.

[75] P. Czarnul, J. Kuchta, and M. Matuszek. Parallel Computations in the Volunteer Based Comcute System. In Springer-Verlag, editor, *Proc. of Parallel Processing and Applied Mathematics (PPAM)*, volume LNCS 8384, Poland, Sept. 2013.

[76] P. Czarnul and J. Kurylowicz. Automatic Conversion of Legacy Applications into Services in BeesyCluster. In *Information Technology (ICIT), 2010 2nd International Conference on*, pages 21–24, 2010.

[77] P. Czarnul, M. Matuszek, M. Wojcik, and K. Zalewski. BeesyBees: Agent-Based, Adaptive & Learning Workflow Execution Module for BeesyCluster. In *Information Technology (ICIT), 2010 2nd International Conference on*, pages 139–142, 2010.

[78] P. Czarnul, M.R. Matuszek, M. Wójcik, and K. Zalewski. BeesyBees: A Mobile Agent-Based Middleware for a Reliable and Secure Execution of Service-Based Workflow Applications in BeesyCluster. *Multiagent and Grid Systems*, 7(6):219–241, 2011.

[79] P. Czarnul and P. Rosciszewski. Optimization of Execution Time under Power Consumption Constraints in a Heterogeneous Parallel System with GPUs and CPUs. In Mainak Chatterjee, Jian-nong Cao, Kishore

Kothapalli, and Sergio Rajsbaum, editors, *Distributed Computing and Networking*, volume 8314 of *Lecture Notes in Computer Science*, pages 66–80. Springer Berlin Heidelberg, 2014.

[80] P. Czarnul, S. Venkatasubramanian, C.D. Sarris, S. Hung, D. Chun, K. Tomko, E.S. Davidson, L.P.B. Katehi, and B. Perlman. Locality Enhancement and Parallelization of an FDTD Simulation. In *Proceedings of the 2001 DoD/HPCMO Users Conference*, U.S.A., 2001.

[81] P. Czarnul and M. Wójcik. Dynamic Compatibility Matching of Services for Distributed Workflow Execution. In Roman Wyrzykowski, Jack Dongarra, Konrad Karczewski, and Jerzy Wasniewski, editors, *PPAM (2)*, volume 7204 of *Lecture Notes in Computer Science*, pages 151–160. Springer, 2011.

[82] D. Martin, M. Burstein, J. Hobbs, O. Lassila, D. McDermott, S. McIlraith, S. Narayanan, M. Paolucci, B. Parsia, T. Payne, E. Sirin, N. Srinivasan, and K. Sycara. *OWL-S: Semantic Markup for Web Services*. France Telecom, Maryland Information and Network Dynamics Lab at the University of Maryland, National Institute of Standards and Technology (NIST), Network Inference, Nokia, SRI International, Stanford University, Toshiba Corporation, and University of Southampton, November 2004. W3C Member Submission, http://www.w3.org/Submission/OWL-S/.

[83] E. Deelman, J. Blythe, Y. Gil, C. Kesselman, G. Mehta, S. Patil, M. Su, K. Vahi, and M. Livny. Pegasus: Mapping Scientific Workflows onto the Grid. In *Across Grids Conference*, Nicosia, Cyprus, 2004. http://pegasus.isi.edu.

[84] E. Deelman, G. Singha, M. Sua, J. Blythea, Y. Gila, C. Kesselmana, G. Mehtaa, K. Vahia, G.B. Berrimanb, J. Goodb, A. Laityb, J.C. Jacob, and D.S. Katzc. Pegasus: A Framework for Mapping Complex Scientific Workflows onto Distributed Systems. *Scientific Programming*, 13, 2005. IOS Press.

[85] B. Dolan. Which Sensors Are Coming to Your Next Smartphone?, May 2011. mobihealthnews, http://mobihealthnews.com/11006/which-sensors-are-coming-to-your-next-smartphone/.

[86] J. Duato, A.J. Pena, F. Silla, R. Mayo, and E.S. Quintana-Orti. rCUDA: Reducing the Number of GPU-Based Accelerators in High Performance Clusters. In *High Performance Computing and Simulation (HPCS), 2010 International Conference on*, pages 224–231, June 2010.

[87] F. Dvorak, D. Kouril, A. Krenek, L. Matyska, M. Mulac, J. Pospisil, M. Ruda, Z. Salvet, J. Sitera, and M. Vocu. gLite Job Provenance. In L. Moreau and I. Foster, editors, *Provenance and Annotation of*

Data, volume 4145 of *Lecture Notes in Computer Science*, pages 246–253. Springer Berlin Heidelberg, 2006.

[88] T. Erl. *Service-Oriented Architecture: Concepts, Technology, and Design*. Prentice Hall PTR, Upper Saddle River, NJ, USA, 2005.

[89] T. Erl, B. Carlyle, C. Pautasso, and R. Balasubramanian. *SOA with REST: Principles, Patterns & Constraints for Building Enterprise Solutions with REST*. Prentice Hall, first edition, August 2012. ISBN: 978-0137012510.

[90] Eucalyptus Systems. *Eucalyptus 4.0.2 User Guide*, November 2014. https://www.eucalyptus.com/docs/eucalyptus/4.0.2/user-guide-4.0.2.pdf.

[91] T. Fahringer, Jun Qin, and S. Hainzer. Specification of Grid Workflow Applications with AGWL: An Abstract Grid Workflow Language. In *Cluster Computing and the Grid, 2005. CCGrid 2005. IEEE International Symposium on*, volume 2, pages 676–685 Vol. 2, May 2005.

[92] R.T. Fielding. *REST: Architectural Styles and the Design of Network-based Software Architectures*. Doctoral dissertation, University of California, Irvine, 2000.

[93] D. Flanagan, J. Farley, W. Crawford, and K. Magnusson. *Java Enterprise in a Nutshell*. O'Reilly and Associates, 1999.

[94] V. Floros, C. Markou, and N. Mastroyiannopoulos. *LCFGng Cluster Installation Guide Version 2.0 (LCG-2)*, July 2004.

[95] Jos A. B. Fortes. Sky Computing: When Multiple Clouds Become One. In *10th IEEE/ACM International Conference on Cluster, Cloud and Grid Computing, CCGrid 2010, 17-20 May 2010, Melbourne, Victoria, Australia*, page 4. IEEE, 2010.

[96] I. Foster and C. Kesselman, editors. *The Grid: Blueprint for a New Computing Infrastructure*. Morgan Kaufmann, August 1998. ISBN 1558604758.

[97] I. Foster, C. Kesselman, J. Nick, and S. Tuecke. The Physiology of the Grid: An Open Grid Services Architecture for Distributed Systems Integration. In *Open Grid Service Infrastructure WG*, June 22 2002. Global Grid Forum, http://www.globus.org/research/papers/ogsa.pdf.

[98] I. Foster, C. Kesselman, and S. Tuecke. The Anatomy of the Grid: Enabling Scalable Virtual Organizations. *International Journal of High Performance Computing Applications*, 15(3):200–222, 2001. http://www.globus.org/research/papers/anatomy.pdf.

[99] The Apache Software Foundation. jUDDI. An Open Source Implementation of OASIS's UDDI v3 Specification. http://juddi.apache.org.

[100] M. Fraczak. Strategies for Dynamic Load Balancing, Replication of Processes and Data in Internet Systems. Master's thesis, Gdansk University of Technology, 2005.

[101] D. Franz, J. Tao, H. Marten, and A. Streit. A Workflow Engine for Computing Clouds. In *CLOUD COMPUTING 2011: The Second International Conference on Cloud Computing, GRIDs, and Virtualization*. ISBN: 978-1-61208-153-3.

[102] R.M. Freund. Polynomial-time Algorithms for Linear Programming Based Only on Primal Scaling and Projected Gradients of a Potential Function. *Math. Program.*, 51:203–222, 1991.

[103] T. Glatard, J. Montagnat, D. Lingrand, and X. Pennec. Flexible and Efficient Workflow Deployment of Data-Intensive Applications on Grids with MOTEUR. *Int. J. High Perform. Comput. Appl.*, 22:347–360, August 2008.

[104] S. Gogouvitis, K. Konstanteli, D. Kyriazis, G. Katsaros, T. Cucinotta, and M. Boniface. *Achieving Real-Time in Distributed Computing: From Grids to Clouds*, chapter Workflow Management Systems in Distributed Environments, pages 115–132. IGI Global, 2012. 10.4018/978-1-60960-827-9.ch007.

[105] J. Gomes, M. David, J. Martins, L. Bernardo, A. García, M. Hardt, H. Kornmayer, J. Marco, R. Marco, D. Rodríguez, I. Diaz, D. Cano, J. Salt, S. Gonzalez, J. Sánchez, F. Fassi, V. Lara, P. Nyczyk, P. Lason, A. Ozieblo, P. Wolniewicz, M. Bluj, K. Nawrocki, A. Padee, W. Wislicki, C. Fernández, J. Fontán, Y. Cotronis, E. Floros, G. Tsouloupas, W. Xing, M. Dikaiakos, J. Astalos, B. Coghlan, E. Heymann, M. Senar, C. Kanellopoulos, A. Ramos, and D. Groen. Experience with the International Testbed in the CrossGrid Project. In *Proceedings of the 2005 European Conference on Advances in Grid Computing*, EGC'05, pages 98–110, Berlin, Heidelberg, 2005. Springer-Verlag.

[106] Gridbus Project: Workflow Language. xWFL2.0. http://gridbus.cs.mu.oz.au/workflow/2.0beta/docs/xwfl2.pdf.

[107] W. Gropp, E. Lusk, and A. Skjellum. *Using MPI: Portable Parallel Programming with the Message Passing Interface*, 2nd edition. MIT Press, Cambridge, MA, 1999.

[108] W. Gropp, E. Lusk, and R. Thakur. *Using MPI-2: Advanced Features of the Message-Passing Interface*. MIT Press, Cambridge, MA, 1999.

[109] G. Hackmann, M. Haitjema, C. Gill, and G. Roman. Sliver: A BPEL Workflow Process Execution Engine for Mobile Devices. In *Proceedings of 4th International Conference on Service Oriented Computing (IC-SOC)*, pages 503–508. Springer Verlag, 2006.

[110] M. Hadley. *Web Application Description Language.* Sun Microsystems, Inc, August 2009. W3C Member Submission, `http://www.w3.org/Submission/wadl/`.

[111] Mark D. Hansen. *SOA Using Java Web Services.* Prentice Hall, May 2007. ISBN: 978-0130449689.

[112] Y. He and C. Ding. Hybrid OpenMP and MPI: Programming and Tuning. In *NUG2004.* Lawrence Berkeley National Laboratory, June 2004.

[113] J. Heimbuch. Android App Measures Air Pollution Using Cell Phone's Camera, September 2010. treehugger, Technology / Clean Technology, `http://www.treehugger.com/clean-technology/android-app-mea` `sures-air-pollution-using-cell-phones-camera.html`.

[114] E. Hellman. *Android Programming: Pushing the Limits.* Wiley, 2013. ISBN: 978-1118717370.

[115] M. Henning. Binding, Migration, and Scalability in CORBA. *Communications of the ACM*, 41(10):62–71, October 1998.

[116] C. Hoffa, G. Mehta, T. Freeman, E. Deelman, K. Keahey, B. Berriman, and J. Good. On the Use of Cloud Computing for Scientific Workflows. In *IEEE Fourth International Conference on eScience '08*, pages 640–645, 2008. doi: 10.1109/eScience.2008.167.

[117] J. Hunter and W. Crawford. *Java Servlet Programming.* O'Reilly, second edition, April 2001.

[118] JADE Board. JADE Web Services Integration Gateway (WSIG) Guide, March 2013. version 3.0, `http://jade.tilab.com/doc/` `tutorials/WSIG_Guide.pdf`, Telecom Italia.

[119] M. Jankowski, P. Wolniewicz, and N. Meyer. Practical Experiences with User Account Management in Clusterix, November 2005. Cracow Grid Workshop, `http://vus.psnc.pl/articles/cgw-05.pdf`.

[120] E. Jendrock, I. Evans, D. Gollapudi, K. Haase, and C. Srivathsa. *The Java EE 6 Tutorial: Basic Concepts.* Prentice Hall Press, Upper Saddle River, NJ, USA, 4th edition, 2010.

[121] M.B. Juric. *Business Process Execution Language for Web Services BPEL and BPEL4WS 2nd Edition.* Packt Publishing, 2006.

[122] G. Juve, A. Chervenak, E. Deelman, S. Bharathi, G. Mehta, and K. Vahi. Characterizing and Profiling Scientific Workflows. *Future Gener. Comput. Syst.*, 29(3):682–692, March 2013.

[123] G. Juve and E. Deelman. *Grids, Clouds and Virtualization*, chapter Scientific Workflows in the Cloud, pages 71–91. Springer, 2010.

[124] K. Kaczmarczyk. A BeesyCluster Module for Discovering Resources, Compilation and Execution of Programs. Master's thesis, Gdansk University of Technology, Faculty of Electronics, Telecommunications and Informatics, 2012.

[125] N. Karmarkar. A New Polynomial-Time Algorithm for Linear Programming. *Combinatorica*, 4(4):373–395, 1984.

[126] N.T. Karonis, B. Toonen, and I. Foster. MPICH-G2: A Grid-Enabled Implementation of the Message Passing Interface. *Journal of Parallel and Distributed Computing*, 63(5):551–563, 2003. Special Issue on Computational Grids.

[127] K. Keahey, M.O. Tsugawa, A.M. Matsunaga, and J.A.B. Fortes. Sky Computing. *IEEE Internet Computing*, 13(5):43–51, 2009.

[128] Rainer Keller and Matthias Mller. The Grid-Computing Library PACX-MPI: Extending MPI for Computational Grids. http://www.hlrs.de/organization/av/amt/research/pacx-mpi/.

[129] D.B. Kirk and W.W. Hwu. *Programming Massively Parallel Processors, Second Edition: A Hands-on Approach*. Morgan Kaufmann, 2012. ISBN-13: 978-0124159921.

[130] J. Kitowski, M. Turala, K. Wiatr, and L. Dutka. PL-Grid: Foundations and Perspectives of National Computing Infrastructure. In Marian Bubak, Tomasz Szepieniec, and Kazimierz Wiatr, editors, *Building a National Distributed e-Infrastructure: PL-Grid*, volume 7136 of *Lecture Notes in Computer Science*, pages 1–14. Springer Berlin Heidelberg, 2012.

[131] P. Kopta, T. Kuczynski, and R. Wyrzykowski. Grid Access and User Interface in CLUSTERIX Project. In Roman Wyrzykowski, Jack Dongarra, Norbert Meyer, and Jerzy Wasniewski, editors, *PPAM*, volume 3911 of *Lecture Notes in Computer Science*, pages 249–257. Springer, 2005.

[132] H. Krawczyk, R. Knopa, and J. Proficz. Basic Management Strategies on KASKADA Platform. In *EUROCON*, pages 1–4. IEEE, 2011.

[133] H. Krawczyk and J. Proficz. KASKADA: Multimedia Processing Platform Architecture. In George A. Tsihrintzis and Maria Virvou, editors,

SIGMAP 2010: Proceedings of the International Conference on Signal Processing and Multimedia Applications, Athens, Greece, July 26–28, 2010, SIGMAP is part of ICETE - The International Joint Conference on e-Business and Telecommunications, pages 26–31. SciTePress, 2010.

[134] L. Kuczynski, K. Karczewski, and R. Wyrzykowski. CLUSTERIX Data Management System and Its Integration with Applications. In Roman Wyrzykowski, Jack Dongarra, Norbert Meyer, and Jerzy Wasniewski, editors, *PPAM*, volume 3911 of *Lecture Notes in Computer Science*, pages 240–248. Springer, 2005.

[135] B.S. Kumar, G. Paliwal, M. Raghav, and S. Nair. Aneka as PaaS (Cloud Computing). *Journal of Computing Technologies*, 2012. ISSN 2278-3814.

[136] M. Kupczyk, R. Lichwala, N. Meyer, B. Palak, M. Plociennik, M. Stroinski, and P. Wolniewicz. The Migrating Desktop as a GUI Framework for the "Applications on Demand" Concept. In Marian Bubak, G. Dick van Albada, Peter M. A. Sloot, and Jack Dongarra, editors, *International Conference on Computational Science*, volume 3036 of *Lecture Notes in Computer Science*, pages 91–98. Springer, 2004.

[137] Large Scale Distributed Information Systems, Department of Computer Science The University of Georgia. METEOR-S: Semantic Web Services and Processes. Applying Semantics in Annotation, Quality of Service, Discovery, Composition, Execution, 2005. http://lsdis.cs.uga.edu/projects/meteor-s/.

[138] E. Laure, C. Gr, S. Fisher, A. Frohner, P. Kunszt, A. Krenek, O. Mulmo, F. Pacini, F. Prelz, J. White, M. Barroso, P. Buncic, R. Byrom, L. Cornwall, M. Craig, A. Di Meglio, A. Djaoui, F. Giacomini, J. Hahkala, F. Hemmer, S. Hicks, A. Edlund, A. Maraschini, R. Middleton, M. Sgaravatto, M. Steenbakkers, J. Walk, and A. Wilson. Programming the Grid with gLite. In *Computational Methods in Science and Technology*, 12(1), 33–45 (2006), Scientific Publishers.

[139] B. Lesyng, P. Bala, and D.W. Erwin. EUROGRID–European computational grid testbed. *J. Parallel Distrib. Comput.*, 63(5):590–596, 2003.

[140] J. Li, Y. Bu, S. Chen, X. Tao, and J. Lu. FollowMe: on Research of Pluggable Infrastructure for Context-Awareness. In *Advanced Information Networking and Applications, 2006. AINA 2006. 20th International Conference on*, volume 1, April 2006.

[141] M. Li and M.A. Baker. *The Grid: Core Technologies*. Wiley, 2005.

[142] B. Linke, R. Giegerich, and A. Goesmann. Conveyor: a Workflow Engine for Bioinformatic Analyses. *Bioinformatics*, 27(7):903–11, 2011.

[143] A. Louis. ESB Topology Alternatives, May 2008. InfoQueue, `http://www.infoq.com/articles/louis-esb-topologies`.

[144] J. Lu and L. Chen. An Architecture for Building User-Driven Web Tasks via Web Services. In *Third International Conference, E-Commerce and Web Technologies EC-Web 2002, Proceedings*, number 2455 in Lecture Notes in Computer Science, pages 77–86, Aix-en-Provence, France, September 2–6 2002. Springer-Verlag.

[145] R. Lucchi, M. Millot, and C. Elfers. Resource Oriented Architecture and REST. Assessment of Impact and Advantages on INSPIRE. Technical Report EUR 23397 EN - 2008, European Commission Joint Research Centre, Institute for Environment and Sustainability, Con terra - Gesellschaft fr Angewandte Informationstechnologie mbH, 2008. DOI 10.2788/80035, `http://inspire.ec.europa.eu/reports/ImplementingRules/netwo rk/Resource_orientated_architecture_and_REST.pdf`.

[146] B. Ludascher, I. Altintas, C. Berkley, D. Higgins, E. Jaeger-Frank, M. Jones, E. Lee, J. Tao, and Y. Zhao. Scientific Workflow Management and the Kepler System. *Concurrency and Computation: Practice & Experience, Special Issue on Scientific Workflows*, 2005.

[147] R. Ma, Y. Wu, X. Meng, S. Liu, and Li P. Grid-Enabled Workflow Management System Based on BPEL. *Int. J. High Perform. Comput. Appl.*, 22(3):238–249, 2008.

[148] M. Maio, M. Salatino, and E. Aliverti. *jBPM6 Developer Guide*. Packt Publishing, August 2014. ISBN: 9781783286614.

[149] N. Maisonneuve, M. Stevens, and L. Steels. Measure and Map Noise Pollution with Your Mobile Phone. In *Proceedings of the DIY for CHI Workshop Held on April 5, 2009 at CHI 2009, the 27th Annual CHI Conference on Human Factors in Computing Systems*, pages 78–82, Boston, MA, USA, April 2009.

[150] S. Majithia, M.S. Shields, I.J. Taylor, and I. Wang. Triana: A Graphical Web Service Composition and Execution Toolkit. In *IEEE International Conference on Web Services (ICWS'04)*, pages 512–524. IEEE Computer Society, 2004.

[151] L. Mandel. *Describe REST Web Services with WSDL 2.0. A How-to Guide*. IBM, May 2008. developerWorks, `http://www.ibm.com/developerworks/webservices/library/ws-re stwsdl/ws-restwsdl-pdf.pdf`.

[152] P. Marcotte, G. Savard, and A. Schoeb. A Hybrid Approach to the Solution of a Pricing Model with Continuous Demand Segmentation. *EURO Journal on Computational Optimization*, 1(1-2):117–142, Jan 2013.

[153] V. Mayer-Schnberger and K. Cukier. *Big Data a Revolution That Will Transform How We Live, Work, and Think*. Eamon Dolan/Mariner Books, 2014. ISBN: 978-0544227750.

[154] P. Mohan, V.N. Padmanabhan, and R. Ramjee. Nericell: Rich Monitoring of Road and Traffic Conditions Using Mobile Smartphones. In *Proceedings of the 6th ACM Conference on Embedded Network Sensor Systems*, SenSys '08, pages 323–336, New York, NY, USA, 2008. ACM.

[155] G. Moro, S. Bergamaschi, and K. Aberer, editors. *Agents and Peer-to-Peer Computing*, volume 3601 of *Lecture Notes in Computer Science, Lecture Notes in Artificial Intelligence*, New York, 2004. http://www.springer.com/computer/ai/book/978-3-540-29755-0.

[156] J. Murty. *Programming Amazon Web Services: S3, EC2, SQS, FPS, and SimpleDB*. O'Reilly Media, first edition, April 2008. ISBN: 978-0596515812.

[157] J.C. Nash. The (Dantzig) Simplex Method for Linear Programming. In *Computing in Science and Eng.*, volume 2, pages 29–31. IEEE Educational Activities Department, Piscataway, NJ, USA, 2000.

[158] E. Ng, S. Tan, and J.G. Guzman. Road Traffic Monitoring Using a Wireless Vehicle Sensor Network. In *Intelligent Signal Processing and Communications Systems, 2008. ISPACS 2008. International Symposium on*, pages 1–4, Feb 2009.

[159] R. Nixon. *Learning PHP, MySQL, JavaScript, CSS & HTML5: A Step-by-Step Guide to Creating Dynamic Websites*. O'Reilly Media, third edition, June 2014. ISBN: 978-1491949467.

[160] R.P. Norris. Data Challenges for Next-Generation Radio Telescopes. In *Proceedings of the 2010 Sixth IEEE International Conference on e-Science Workshops*, E-SCIENCEW '10, pages 21–24, Washington, DC, USA, 2010. IEEE Computer Society.

[161] NVIDIA. CUDA Programming Guide. version 5.5, http://docs.nvidia.com/cuda/cuda-c-programming-guide/.

[162] University of California. BOINC. http://boinc.berkeley.edu/.

[163] University of California. Data Server Protocol. BOINC. http://boinc.berkeley.edu/trac/wiki/FileUpload.

[164] OMG (Object Management Group). *Common Object Request Broker Architecture (CORBA/IIOP): Core Specification*, v. 3.0.2 edition, December 2002. http://www.omg.org/spec/CORBA/3.0.2/PDF.

[165] OMG (Object Management Group). Business Process Model and Notation (BPMN), Version 2.0, January 2011. `http://www.omg.org/spec/BPMN/2.0`.

[166] R. Orfali, D. Harkey, and J. Edwards. *The Essential Client/Server Survival Guide, 3rd Edition*. Wiley, 1999.

[167] S. Ostermann, R. Prodan, and T. Fahringer. Resource Management for Hybrid Grid and Cloud Computing. In Nick Antonopoulos and Lee Gillam, editors, *Cloud Computing, Computer Communications and Networks*, pages 179–194. Springer London, 2010.

[168] J. Pai. JBoss Application Server 7.1 Documentation. Developer Guide. EJB Invocations from a Remote Client Using JNDI, November 2012. `https://docs.jboss.org/author/display/AS71/EJB+invocations+from+a+remote+client+using+JNDI`.

[169] D. Panda, R. Rahman, and D. Lane. *EJB 3 in Action*. Manning Publications, 2007. ISBN: 978 1933988344.

[170] S. Pandey, D. Karunamoorthy, and R. Buyya. *Workflow Engine for Clouds*, chapter Cloud Computing: Principles and Paradigms. Wiley Press, New York, USA, 2011. ISBN-13: 978-0470887998.

[171] A. Papaspyrou. *Instant Google Compute Engine*. Packt Publishing, September 2013. ISBN: 978-1849697002.

[172] C. Patel, K. Supekar, and Y. Lee. A QoS Oriented Framework for Adaptive Management of Web Service Based Workflows. In *Proceedings of the 14th International Database and Expert Systems Applications Conference (DEXA 2003)*, LNCS, pages 826–835, Prague, Czech Republic, September 2003.

[173] S. Poduri, A. Nimkar, and G. Sukhatme. Visibility Monitoring Using Mobile Phones, December 2009. Department of Computer Science, University of Southern California, `http://robotics.usc.edu/mobilesensing/visibility/MobileAirQualitySensing.pdf`.

[174] S. Pousty and K. Miller. *Getting Started with OpenShift*. O'Reilly Media, June 2014. ISBN: 978-1491900475.

[175] A. Puder and K. Romer. *MICO: An Open Source CORBA Implementation*. Morgan Kaufmann Publishers, 2000. ISBN 978-1558606661.

[176] P. Rajasekaran, J. Miller, K. Verma, and A. Sheth. Enhancing Web Services Description and Discovery to Facilitate Composition. In *Proceedings of the First International Conference on Semantic Web Services and Web Process Composition*, SWSWPC'04, pages 55–68, Berlin, Heidelberg, 2005. Springer-Verlag.

[177] The Regents of the University of California, through Lawrence Berkeley National Laboratory. *Berkeley Lab Checkpoint/Restart (BLCR) User's Guide*, 2012. http://crd.lbl.gov/groups-depts/ftg/projects/current-projects/BLCR.

[178] R. Reyes, I. López-Rodríguez, J.J. Fumero, and F. de Sande. A Preliminary Evaluation of OpenACC Implementations. *The Journal of Supercomputing*, 65(3):1063–1075, 2013.

[179] J. Rhoton, J. De Clercq, and F. Novak. *OpenStack Cloud Computing: Architecture Guide*. Recursive Press, 2014. ISBN: 978-0956355683.

[180] L. Richardson and S. Ruby. *Restful Web Services*. O'Reilly, first edition, 2007.

[181] J.W. Romein. An Efficient Work-Distribution Strategy for Gridding Radio-Telescope Data on Gpus. In *Proceedings of the 26th ACM international conference on Supercomputing*, ICS '12, pages 321–330, New York, NY, USA, 2012. ACM.

[182] W. Roshen. Enterprise Service Bus with USB-like Universal Ports. In *Web Services (ECOWS), 2011 Ninth IEEE European Conference on*, pages 177–183, Sept 2011.

[183] A.L. Rubinger and B. Burke. *Enterprise JavaBeans 3.1*. O'Reilly Media, sixth edition, September 2010. ISBN: 978-0596158026.

[184] J. Russell and R. Cohn. *Dcraw*. Book on Demand, 2012.

[185] S. Graham, D. Davis, S. Simeonov, G. Daniels, P. Brittenham, Y. Nakamura, P. Fremantle, D. Koenig, and C. Zentner. *Building Web Services with Java: Making Sense of XML, SOAP, WSDL and UDDI*. SAMS Publishing, 2nd edition, 2004. ISBN: 978-0672326417.

[186] R. Sakellariou, H. Zhao, E. Tsiakkouri, and M.D. Dikaiakos. Scheduling Workflows with Budget Constraints. In *Integrated Research in Grid Computing*, S. Gorlatch and M. Danelutto, Eds.: CoreGrid series. Springer-Verlag, 2007.

[187] Rizos Sakellariou, Henan Zhao, and Ewa Deelman. Mapping Workflows on Grid Resources: Experiments with the Montage Workflow. In Frdric Desprez, Vladimir Getov, Thierry Priol, and Ramin Yahyapour, editors, *Grids, P2P and Services Computing*, pages 119–132. Springer US, 2010.

[188] J. Sanders and E. Kandrot. *CUDA by Example: An Introduction to General-Purpose GPU Programming*. Addison-Wesley Professional, 2010. ISBN-13: 978-0131387683.

[189] D. Sarna. *Implementing and Developing Cloud Computing Applications*. CRC Press, Taylor & Francis Group, 2011. ISBN 978-1-4398-3082-6.

[190] C.D. Sarris, P. Czarnul, S. Venkatasubramanian, W. Thiel, K. Tomko, L.P.B. Katehi, and B.S. Perlman. Mixed Electromagnetic-Circuit Modeling and Parallelization for Rigorous Characterization of Cosite Interference in Wireless Communication Channels. *DoD/HPCMO Users Conference*, U.S.A., 2002.

[191] C.D. Sarris, K. Tomko, P. Czarnul, S. Hung, R.L. Robertson, D. Chun, E.S. Davidson, and L.P.B. Katehi. Multiresolution Time Domain Modeling for Large Scale Wireless Communication Problems. In *Proceedings of the 2001 IEEE AP-S International Symposium on Antennas and Propagation*, volume 3, pages 557–560, 2001.

[192] P. Schwan. Lustre: Building a File System for 1,000-Node Clusters. *Proceedings of the Linux Symposium. July 23th–26th*, Ottawa, Ontario, Canada, 2003.

[193] B. Schwartz, P. Zaitsev, and V. Tkachenko. *High Performance MySQL: Optimization, Backups, and Replication: Covers Version 5.5*, Third Edition. O'Reilly, 2012.

[194] K. Seetharaman. The CORBA Connection. *Communications of the ACM*, 41(10):34–36, October 1998.

[195] C.R. Senna, L.F. Bittencourt, and E.R.M. Madeira. An Environment for Evaluation and Testing of Service Workflow Schedulers in Clouds. In *High Performance Computing and Simulation (HPCS), 2011 International Conference on*, pages 301–307, July 2011.

[196] P.C.-Y. Sheu, H. Yu, C.V. Ramamoorthy, A.K. Joshi, and L.A. Zadeh. *Semantic Computing*. Wiley, Hoboken, NJ, 2010.

[197] V. Silva. *Grid Computing for Developers*. Charles River Media, Hingham, Mass, 2005.

[198] N. Srinivasan, M. Paolucci, and K. Sycara. Adding OWL-S to UDDI, Implementation and Throughput. In *First International Workshop on Semantic Web Services and Web Process Composition (SWSWPC 2004)*, San Diego, California, USA, 2004.

[199] W. Stallings. *Data and Computer Communications*. PEARSON, Prentice Hall, eighth edition, 2007. ISBN: 0-13-243310-9.

[200] M. Still. *The Definitive Guide to ImageMagick*. Apress, December 2005. ISBN 9781590595909.

[201] A. Streit, P. Bala, A. Beck-Ratzka, K. Benedyczak, S. Bergmann, R. Breu, J.M. Daivandy, B. Demuth, A. Eifer, A. Giesler, B. Hagemeier, S. Holl, V. Huber, N. Lamla, D. Mallmann, A.S. Memon, M.S. Memon, M. Rambadt, M. Riedel, M. Romberg, B. Schuller, T. Schlauch,

A. Schreiber, T. Soddemann, and W. Ziegler. UNICORE 6 - Recent and Future Advancements. *Annales des Telecommunications*, 65(11–12):757–762, 2010.

[202] R.E. Sward and K.J. Whitacre. A Multi-Language Service-Oriented Architecture Using an Enterprise Service Bus. In *Proceedings of the 2008 ACM Annual International Conference on SIGAda Annual International Conference*, SIGAda '08, pages 85–90, New York, NY, USA, 2008. ACM.

[203] M.M. Syslo, N. Deo, and J.S. Kowalik. *Discrete Optimization Algorithms*. Prentice-Hall, 1983.

[204] I.J. Taylor, E. Deelman, D.B. Gannon, and M. Shields, editors. *Workflows for e-Science. Scientific Workflows for Grids*. Springer, 2007. ISBN 978-1-84628-519-6.

[205] J. Evdemon Editor(s): A. Alves, A. Arkin, S. Askary, C. Barreto, B. Bloch, F. Curbera, M. Ford, Y. Goland, A. Guzar, N. Kartha, C.K. Liu, R. Khalaf, D. Knig, M. Marin, V. Mehta, S. Thatte, D. van der Rijn, P. Yendluri, and A. Yiu, Technical Committee: OASIS Web Services Business Process Execution Language (WSBPEL) TC, Chair(s): D. Jordan. Web Services Business Process Execution Language Version 2.0, OASIS Standard, April 2007.

[206] The Globus Alliance. *GT 6.0 GRAM5: User's Guide*. `http://toolkit.globus.org/toolkit/docs/latest-stable/gram5/` user.

[207] The Globus Alliance. *GT 6.0 GridFTP: User's Guide*. `http://toolkit.globus.org/toolkit/docs/latest-stable/gridftp/user`.

[208] The Globus Alliance. *Security: GSI C User's Guide*. `http://toolkit.globus.org/toolkit/docs/latest-stable/gsic/user`.

[209] The Legion Group. *Legion 1.8. System Administrator Manual*. University of Virginia, VA, USA. `http://legion.virginia.edu/documentation/sysadmin_1.8.pdf`.

[210] The OpenMP Architecture Review Board. OpenMP. The OpenMP API Specification for Parallel Programming. `http://openmp.org/wp/`.

[211] J. Vockler, G. Juve, E. Deelman, M. Rynge, and G. Berriman. Experiences Using Cloud Computing for a Scientific Workflow Application. In *Proceedings of 2nd Workshop on Scientific Cloud Computing (ScienceCloud 2011)*, 2011.

[212] W3C (MIT, ERCIM, Keio). *SOAP Version 1.2*, April 2007. W3C Recommendation (Second Edition), `http://www.w3.org/TR/soap/`.

[213] W3C (MIT, ERCIM, Keio). *Web Services Description Language (WSDL) Version 2.0 Part 1: Core Language*, June 2007. Editors: Roberto Chinnici, Jean-Jacques Moreau, Arthur Ryman, Sanjiva Weerawarana, W3C Recommendation, http://www.w3.org/TR/wsdl20/.

[214] C. Wang, F. Mueller, C. Engelmann, and S.L. Scott. A Job Pause Service under LAM/MPI+BLCR for Transparent Fault Tolerance. In *Parallel and Distributed Processing Symposium, 2007. IPDPS 2007. IEEE International*, pages 1–10, 2007.

[215] M. Wieczorek, R. Prodan, and T. Fahringer. Scheduling of Scientific Workflows in the ASKALON Grid Environment. *SIGMOD Rec.*, 34(3):56–62, 2005.

[216] Marek Wieczorek, Andreas Hoheisel, and Radu Prodan. Towards a General Model of the Multi-Criteria Workflow Scheduling on the Grid. *Future Generation Computer Systems*, 25(3):237 – 256, 2009.

[217] S. Wienke, P. Springer, C. Terboven, and D. an Mey. OpenACC: First Experiences with Real-World Applications. In *Proceedings of the 18th International Conference on Parallel Processing*, Euro-Par'12, pages 859–870, Berlin, Heidelberg, 2012. Springer-Verlag.

[218] B. Wilkinson and M. Allen. *Parallel Programming: Techniques and Applications Using Networked Workstations and Parallel Computers*. Prentice Hall, 1999.

[219] R. Wyrzykowski, N. Meyer, T. Olas, L. Kuczynski, B. Ludwiczak, C. Czaplewski, and S. Oldziej. Meta-computations on the CLUSTERIX Grid. In Bo Kgstrm, Erik Elmroth, Jack Dongarra, and Jerzy Wasniewski, editors, *PARA*, volume 4699 of *Lecture Notes in Computer Science*, pages 489–500. Springer, 2006.

[220] R. Wyrzykowski, N. Meyer, and M. Stroinski. Concept and Implementation of CLUSTERIX: National Cluster of Linux Systems. In *6th LCI International Conference on Linux Clusters: The HPC Revolution*, Chapel Hill, NC, USA, 2005.

[221] G.E. Yu, P. Zhao, L. Di, A. Chen, M. Deng, and Y. Bai. BPELPowerA {BPEL} Execution Engine for Geospatial Web Services. *Computers & Geosciences*, 47(0):87–101, 2012. Towards a Geoprocessing Web.

[222] J. Yu and R. Buyya. A Taxonomy of Workflow Management Systems for Grid Computing. *Journal of Grid Computing*, 3(3-4):171–200, September 2005.

[223] J. Yu and R. Buyya. A Budget Constrained Scheduling of Workflow Applications on Utility Grids Using Genetic Algorithms. In *Workshop on Workflows in Support of Large-Scale Science, Proceedings of the 15th*

IEEE International Symposium on High Performance Distributed Computing (HPDC 2006), Paris, France, June 2006.

[224] J. Yu, R. Buyya, and K. Ramamohanarao. *Workflow Scheduling Algorithms for Grid Computing*, chapter Metaheuristics for Scheduling in Distributed Computing Environments. Springer, 2008.

[225] J. Yu, R. Buyya, and C. Tham. Cost-Based Scheduling of Workflow Applications on Utility Grids. In *Proceedings of the 1st IEEE International Conference on e-Science and Grid Computing (e-Science 2005), IEEE CS Press*, Melbourne, Australia, December 2005.

[226] L. Zeng, B. Benatallah, M. Dumas, J. Kalagnanam, and Q. Sheng. Quality Driven Web Services Composition. In *Proceedings of WWW 2003*, Budapest, Hungary, May 2003.

[227] L. Zeng, B. Benatallah, A.H.H. Ngu, M. Dumas, J. Kalagnanam, and H. Chang. QoS-Aware Middleware for Web Services Composition. *IEEE Trans. Softw. Eng.*, 30(5):311–327, 2004.

[228] Y. Zheng, M. Zhao, Y. Song, H. Adam, U. Buddemeier, A. Bissacco, F. Brucher, T. Chua, and H. Neven. Tour the World: Building a Web-scale Landmark Recognition Engine. *2013 IEEE Conference on Computer Vision and Pattern Recognition*, 0:1085–1092, 2009.

Index